Enhancing Strategic Planning with Massive Scenario Generation

Theory and Experiments

Paul K. Davis, Steven C. Bankes, Michael Egner

RAND NATIONAL SECURITY RESEARCH DIVISION

This research was conducted within the Intelligence Policy Center (IPC) of the RAND National Security Research Division (NSRD). NSRD conducts research and analysis for the Office of the Secretary of Defense, the Joint Staff, the Unified Commands, the defense agencies, the Department of the Navy, the Marine Corps, the U.S. Coast Guard, the U.S. Intelligence Community, allied foreign governments, and foundations.

Library of Congress Cataloging-in-Publication Data

Davis, Paul K., 1943-
    Enhancing strategic planning with massive scenario generation : theory and experiments / Paul K. Davis, Steven C. Bankes, Michael Egner.
        p. cm.
    Includes bibliographical references.
    ISBN 978-0-8330-4017-6 (pbk. : alk. paper)
    1. Command of troops. 2. Decision making—Methodology. 3. Military planning—Decision making.
    I. Bankes, Steven C. II. Egner, Michael. III. Title.

UB210.D3875 2007
355.6'84—dc22

2007016537

The RAND Corporation is a nonprofit research organization providing objective analysis and effective solutions that address the challenges facing the public and private sectors around the world. RAND's publications do not necessarily reflect the opinions of its research clients and sponsors.

RAND® is a registered trademark.

Published 2007 by the RAND Corporation
1776 Main Street, P.O. Box 2138, Santa Monica, CA 90407-2138
1200 South Hayes Street, Arlington, VA 22202-5050
4570 Fifth Avenue, Suite 600, Pittsburgh, PA 15213-2665
RAND URL: http://www.rand.org/
To order RAND documents or to obtain additional information, contact
Distribution Services: Telephone: (310) 451-7002;
Fax: (310) 451-6915; Email: order@rand.org

Preface

As indicated by the title, this report describes experiments with new methods for strategic planning based on generating a very wide range of futures and then drawing insights from the results. The emphasis is not so much on "massive scenario generation" per se as on thinking broadly and open-mindedly about what may lie ahead. The report is intended primarily for a technical audience, but the summary should be of interest to anyone curious about modern methods for improving strategic planning under uncertainty. Comments are welcome and should be addressed to Paul K. Davis or Steven Bankes at the RAND Corporation. Their e-mail addresses are Paul_Davis@rand.org and Steven_Bankes@rand.org.

This research was conducted within the Intelligence Policy Center of the RAND National Security Research Division (NSRD), which also supported extension of the work and preparation of this report. NSRD conducts research and analysis for the Office of the Secretary of Defense, the Joint Staff, the Unified Combatant Commands, the defense agencies, the Department of the Navy, the Marine Corps, the U.S. Coast Guard, the U.S. Intelligence Community, allied foreign governments, and foundations.

For more information on RAND's Intelligence Policy Center, contact the Director, John Parachini. He can be reached by e-mail at John_Parachini@rand.org; by phone at 703-413-1100, extension 5579; or by mail at the RAND Corporation, 1200 South Hayes Street, Arlington, Virginia 22202-5050. More information about RAND is available at www.rand.org.

Contents

Figures

Tables

Summary

A general problem in strategic planning is that planning should be informed by a sense of what may lie ahead—not only the obvious possibilities that are already in mind, but also a myriad of other possibilities that may not have been recognized, or at least may not have been understood and taken seriously. Various techniques, including brainstorming sessions and human gaming, have long been used to raise effective awareness of possibilities. Model-based analysis has also been used and, over the last decade, has been used to explore sizable numbers of possible scenarios. This report extends that research markedly, with what can be called massive scenario generation (MSG). Doing MSG well, as it turns out, requires a conceptual framework, a great deal of technical and analytical thinking, and—as in other forms of scientific inquiry—experimentation. This report describes our recent progress on both theory and experimentation. In particular, it suggests ways to combine the virtues of human-intensive and model-intensive exploration of "the possibility space."

Vision: What MSG Might Accomplish

MSG has the potential to improve planning in at least three ways: intellectual, pragmatic, and experiential. Intellectually, MSG should expand the scope of what is recognized as possible developments, provide an understanding of how those developments might come about, and help identify aspects of the world that should be studied more carefully, tested, or monitored. Pragmatically, MSG should assist planners by enriching their mental library of the "patterns" used to guide reasoning and action at the time of crisis or decision, and it should also help them identify anomalous situations requiring unusual actions. MSG should facilitate development of flexible, adaptive, and robust (FAR) strategies that are better able to deal with uncertainty and surprises than strategies based on narrow assumptions. As a practical matter, MSG should also identify crucial issues worthy of testing or experimentation in games or other venues. And, in some cases, it should suggest ways to design mission rehearsals so as to better prepare those executing operational missions. At the experiential level, if MSG can be built into training, education, research, and socialization exercises, it should leave participants with a wider and better sense of the possible, while developing skill at problem-solving in situations other than those of the "best estimate."

The Challenge and Related Needs

It is one thing to have a vision of what MSG might be good for; it is quite another to define what it means, what is necessary for it to be effective, and how it can be accomplished. Such definition will require considerable work over a period of years, but our initial research, described here, provides direction, numerous technical and analytical suggestions, and better-posed problems for that future work.

The Need for Models of a Different Kind

The value of MSG for strategic analysis depends on whether the scenarios that are generated are meaningful and understandable. There is little value in a magical machine that spews out scenarios that are merely descriptions of some possible state of the world; we need to be able to understand how such developments might occur and what their implications might be. In practice, this leads to the need to generate scenarios with a model that can provide the necessary structure and explanation. A dilemma, however, is that models often restrict the scope of thinking—the exact opposite of what is intended here—because they represent particular views of the world and reflect a great many dubious assumptions. Another problem is that in strategic analysis it is often necessary to begin work without the benefit of even a good prior model.

Metrics for Evaluating Methods of MSG

Because of such issues, we suggest that the virtues of a particular approach to MSG can be measured against four metrics: not needing a good initial model; the dimensionality of the possibility space considered; the degree of exploration of that space; and the quality of resulting knowledge.

Two Experiments

With these metrics in mind, we conducted two MSG experiments for contrasting cases. The first case began with a reasonable but untested analytical model, one describing the rise and fall of Islamist extremism in epidemiological terms and relating results to hypothetical policy actions and numerous other parameters. The second case began without an analytical model, but with a thoughtful list (provided by another study) of the conditions that might characterize and distinguish among circumstances at the time of the next nuclear use (NNU). Such a list of conditions might have been developed by, for example, a political scientist writing a thoughtful essay or a strategic-planning exercise in which a number of experts brainstorm about the NNU. The two experiments with MSG therefore covered very different cases, the first having advantages for ultimate knowledge and exploration and the second having the advantage of not requiring an initial model.

In the first experiment, we discovered how inadequate the usual approach to modeling is for the purposes of MSG. The initial analytical model was quite reasonable by most standards, but it omitted discussion of factors that, in the real world, might dominate the problem. This was hardly unusual, since ordinary modeling tends to gravitate toward idealizations, which have many virtues. Nonetheless, in our application, we had to amend the model substantially.

In particular, and despite our aversion to introducing stochastic processes that often serve merely to make models more complicated and data demands more extensive, we concluded that "exogenous" world events, which are arguably somewhat random, could have large effects on the rise or decline of Islamist extremism. Thus, we allowed for such events. We also recognized that well-intended policies to combat extremism are often beset by the possibility of their proving to be counterproductive. That is, in the real world, we often do not even know the *direction* of the arrow in an influence diagram! The result of our enhancements was to construct a highly parameterized model that also allowed for deep uncertainties about external events and when they might occur, and for deep uncertainties about the effectiveness of possible policies. The resulting MSG generated a much richer and more insightful set of possible scenarios than would otherwise have been obtained. Although our analysis was merely illustrative, as part of an experiment on MSG, we concluded that the insights from it were both interesting and nontrivial—enough, certainly, to support the belief that MSG can prove quite fruitful.

Our second experiment, on the NNU, required even more iteration and contemplation because the structure that we began with was "static," a set of possible situational attributes. This proved inadequate to the purposes of MSG, and we concluded that the appropriate approach from such a starting point was to construct quickly the sketch of a dynamic system model, even though not enough time was available to do so well. Once a relevant system-level "influence diagram" had been sketched, we could move to a first-cut dynamic model that was capable of both generating diverse scenarios and providing enough context and history to enable the scenarios to be more like significant causal stories. As in our first experiment, we found ourselves dissatisfied with the initial notions of influence and causality because they were far too certain to be realistic. Thus, we developed techniques that varied the directions and magnitudes of the postulated influences, while also filtering out some of those we considered to be impossible or beyond the pale. Any such filtering, of course, had to be done with caution because of the concern that applying apparently reasonable filtering could in fact eliminate possibilities that should be considered. In any case, having done the first-cut system modeling and introduced the uncertainties of influence and magnitude, we found that MSG produced "data" that included interesting scenarios that would not usually be considered and plausible insights that could affect strategy development. Although we were merely experimenting and would not want to exaggerate the real-world significance or credibility of the experiment's outcome, our conclusion—in contradiction to our initial thinking—was that the method showed significant promise. The key, however, was to recognize the importance of constructing a model early on, even if it could be only at the level of influence diagrams and initial rules of thumb. Taking that step changes the entire direction of scenario generation and provides a core that can be enriched iteratively through hard thinking, brainstorming, gaming, and other mechanisms.

Methods for Interpreting Results of MSG

A major part of our work consisted of experimenting with a variety of methods and tools for interpreting and making sense of the "data" arising from MSG. It is one thing to generate thousands or even tens of thousands of scenarios, but then what? In this study we used four primary methods: (1) ordinary linear sensitivity analysis, (2) a generalization using analyst-inspired "aggregation fragments," (3) some advanced "filtering" methods drawing on data-mining and machine-learning methods, and (4) *motivated metamodeling*. The first three methods were particularly useful for identifying which parameters potentially had the most effect on scenario outcomes, a prerequisite for developing good visualizations. The fourth method involved looking for an analytical "model of the model" (a metamodel) that would provide a relatively simple explanation for scenario outcomes. Motivated metamodeling applies standard statistical machinery to analyze data but starts with a hypothesized analytical structure motivated by an understanding of the subject area.

Tools for Visualizing and Interpreting Results

We used two primary tools, those of the Analytica® modeling system and those of the CARs® system, which can generate scenarios using various models and then help in analysis of the results with many statistical techniques, such as the filters mentioned above. CARs also has good visualization capabilities and can deal with very large numbers of scenarios (we routinely generated tens of thousands). One goal of our work with these tools (primarily CARs) was to find ways to use visualization methods to extract "signal from noise" in analyzing outcomes from MSG. We drew on much past work in doing so, but the challenges in the current effort were new in many respects. As discussed in the text of the report, we were heartened by the results and concluded that the tools have substantial potential. Pursuing that potential will be exciting new research.

Acknowledgments

The authors appreciate reviews by James Dewar of RAND and Paul Bracken of Yale University. They also benefited from discussions of exploratory analysis and exploratory modeling with numerous RAND colleagues over time.

Abbreviations

ABM	agent-based model
ABP	assumption-based planning
CAS	complex adaptive systems
CBP	capabilities-based planning
DPRK	Democratic People's Republic of Korea
EA	exploratory analysis
EBO	effects-based operations
FAR	flexible, adaptive, and robust (as in FAR strategies)
GBN	Global Business Network
IR&D	Internal Research and Development
MRM	multiresolution modeling
MSG	massive scenario generation
NNU	next nuclear use
PRIM	Patient Rule Induction Method
R&D	research and development
RAP	robust adaptive planning
RSAS	RAND Strategy Assessment System

1. Introduction

Objectives

Strategic planning serves many functions. These include conceiving broad strategic options creatively, evaluating and choosing among them, defining in some detail strategies to deal with coordination and integration, implementing those strategies in complicated organizations, and preparing leadership at multiple levels of organization both for as-planned developments and for dealing with potential contingencies.

Some of these functions require focus, detail, consistency, and convergent analysis. Others require creativity and *divergent* thinking. We are concerned here with the more creative functions. Even within those, there is need for a mix of divergent and convergent thinking. In the divergent phase, one wants to consider more than a single, canonical image of the future and, indeed, to consider a broad range of possibilities. These futures may differ, for example, in external considerations (e.g., economic growth, world events), in one's own strategy, and in the strategies of competitors or opponents. Creativity is also desirable in conceiving one's own strategy, as well as in anticipating those of others and plausible external events.[1]

After a period of divergent thinking about the various possibilities, it is necessary to make judgments and decisions—i.e., to converge on a course of action. Both divergent and convergent activities of this type are notoriously difficult.

Against this general background, this report presents a theory of how to use models and computational experiments to help understand the full diversity of possible futures and to draw implications for planning. That is, the intent is to confront uncertainty and suggest strategies to deal with it. The techniques used are what we call *massive scenario generation* (MSG) and *exploratory analysis* (EA).

After a brief review of older methods, we discuss how MSG and advanced methods of EA can contribute. The discussion builds on our past work but goes into considerably more depth on questions such as, How much is enough? in scenario generation, the kinds of models needed to achieve maximum benefit, and the kinds of methods that can be used for convergence.

---

[1] In this report, *plausible* means *possible*, a common usage in strategic planning. This is different from the primary dictionary definition, i.e., *credible* or *likely*. In seeking to anticipate "plausible external events," we essentially mean all external events that are not impossible. We certainly do not mean only events currently thought to be likely.

Divergent Thinking in Strategic Planning

The General Challenge

To appreciate the general challenge, consider first a concrete example: strategic planning at the end of the Cold War, circa 1990. What would come next? Would the Soviet Union collapse but remain intact? Would it disintegrate? Would it fall into civil war? What would happen in Eastern Europe? Would the end of the bipolar era mean a new kind of strife or a unipolar stability of some sort? Analogous but remarkably different questions applied at slices in time such as 1995 or September 12, 2001. What is perhaps most important to note is that *history could have proceeded along any of many different paths.* Inevitability of events is primarily a fiction of after-the-fact historians. That said, planning must proceed if we are to be more than simply passive observers. As became particularly clear in the post–Cold War period, an important part of planning is seeking, where feasible, to *shape* the future environment favorably; an even more important part is preparing for the various possible futures that lie ahead.[2]

Today, the United States is simultaneously in what may be a long war with militant Islamism and associated terrorists and what may be a long-term competition with the ascending powers of Asia. The United States is also greatly concerned about states such as Iran and North Korea, to say nothing of the continuing problems in Iraq. The list of question-mark states goes on, and it should be evident that no one can predict what lies ahead. During the Cold War, conceits about predictability were fairly common (although not among the wise). That is surely not the case today.

The General Technical Challenge

From a technical perspective, a general challenge in the divergent-thinking phase of strategic planning is suggested by Figure 1.1. The notion of the overall figure is that we first open our minds and then make sense of what we learn and converge on insights and perhaps conclusions.[3]

The intention is not simply to move beyond a single, canonical view of the future, but to confront uncertainty as realistically as possible—conceiving the full *"possibility space."* To be sure, even the most heroic efforts are unlikely to be fully successful, as suggested by the difference between the dark and white areas in the middle of Figure 1.1. We aspire, however, to identify as much of the possibility space as possible. We may choose later to dismiss portions of it as insufficiently plausible to worry about or as irrelevant to most planning (e.g., a comet might destroy the earth). Further, we most certainly do not need high levels of detail for all of the points in the possibility space. Nonetheless, we need first to see the possibilities, at least in the large. How do we do so?

---

[2] For examples of the many published uncertainty-sensitive planning efforts over the past 16 years, see the Department of Defense's *Quadrennial Defense Reviews* (Cohen, 1997; Rumsfeld, 2001, 2006), intelligence community documents (National Intelligence Council, 2000, 2004), or various RAND studies (e.g., Davis, 1994b; Wolf et al., 2003).

[3] In practice, the process need not be stepwise as shown. For example, one may jump ahead to imagine disastrous situations and then work *backward* to identify how such situations could arise and what might be done to avoid them. That is part of the methodology of the RAND "Day After" games (Millot, Molander, and Wilson, 1993).

Figure 1.1
Divergence and Convergence

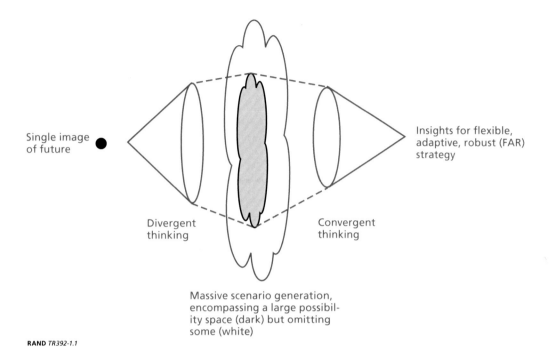

Single image of future

Divergent thinking

Convergent thinking

Insights for flexible, adaptive, robust (FAR) strategy

Massive scenario generation, encompassing a large possibility space (dark) but omitting some (white)

RAND *TR392-1.1*

Scenario-Based Methods and Human Games

The first step in strategic planning's divergent thinking is perhaps the most important: breaking the shackles that bind us to canonical images of the future. The best known planning methods for doing so involve scenarios. The word *scenario* has diverse meanings but is best understood as a postulated sequence of possible events with some degree of internal coherence, i.e., events associated with a "story." Long before the discipline of strategic planning existed, people had learned how to use stories to open minds, break down barriers of certitude, and gain insights from challenges and dilemmas.[4] Scenarios serve a similar purpose.

Scenario-based methods in strategic planning are described in a number of books and have evolved over a half-century since the pioneering work of Herman Kahn,[5] Pierre

---

[4] Parables are examples, some tracing back thousands of years. Many parables deal with real or imagined *past* events (e.g., the parable of King Solomon's tentative ruling that a child should be cut in half to be fair to the two women claiming to be its mother), but they often pose dilemmas, the appreciation of which opens minds. Thus, they serve functions similar in some respects to strategic-planning scenarios.

[5] Kahn's earliest work with scenarios was about nuclear strategy (Kahn, 1966), but later, at his Hudson Institute, he became the foremost futurist on much more general matters. One of his most remarkable books foresaw, during a period of serious economic problems, the coming boom of the 1980s (Kahn, 1983). A recent biography (Ghamari-Tabrizi, 2005) discusses Kahn and his career, although not from a substantive perspective.

Wack (1985), and Peter Schwartz (1995). Schwartz subsequently formed the Global Business Network (GBN). A simple search of the Internet demonstrates how prevalent scenario-based methods are.

One of the most interesting and efficient scenario-based methods is the "Day After" exercise, developed at RAND by Dean Millot, Roger Molander, and Peter Wilson and subsequently applied to quite a number of different problem areas (Millot et al., 1993; Molander et al., 1998; Mussington, 2003). Arguably, "Day After" games are much less about planning per se than about raising consciousness, opening minds, stimulating thought, and kick-starting subsequent planning processes. This functionality is consistent with what some in the business world have also noted, i.e., that a substantial fraction of the value of scenario-based work comes in the first few hours of mind-opening and creativity (van der Werff, 2000).

Human games overlap significantly with scenario-based planning. Often, for example, a game begins with an initiating scenario providing context and "spin" related to the game's purpose. Participants may engage in free play thereafter, which results in a future being played out—perhaps with some branches noted along the way. The particular future is subsequently described as the game's scenario.

Scenario-based planning has long since demonstrated its value, even in empirical work (Schoemaker, 1995), as have games. When led with wisdom and artistry, both methods can be powerful. They also have shortcomings. One of the troubling aspects is that while scenarios and games can open minds relative to the canonical future, they can also trap people in new conceptual structures that are as limiting in their own way as was the original structure. In the worst instances, and despite the admonitions of experts in the use of scenarios and gaming, people may emerge from scenario-based exercises with a sense of inevitability—whether of doom or of glorious success. Even when the intent is to help people to plan under uncertainty, participants may succumb to their instinctual desire to pick a story of the future and then embrace it firmly.[6]

Another criticism is that scenario-based planning often examines only two or three dimensions of uncertainty. The criticism is unfair, because exercise designers examine more dimensions before focusing on those that seem most salient. That, however, is done off-line, and participants may not get the benefit of the broader look. Gaming routinely allows many dimensions of the world to play a role, especially political-military or political-economic gaming, such as the exercises conducted yearly at the Naval War College's Global War Game in Newport, RI.

More troublesome about some of the strategic games and exercises is the fact that the future may be seen as dichotomous (i.e., "things will go either this way or that way"), rather than as more continuously dynamic, with any number of possible branches along the way. And, finally, there is something ad hoc about scenario-based planning: Only some dimensions are focused on, and the reasons for the choice may be neither discussed nor compelling. This

---

[6] This tendency is related to the well known "availability bias" of psychology (Tversky and Kahneman, 1973).

would seem to be a fundamental problem, because scenario-based planning, despite being creative and liberating in some respects, is reductionist in other respects. Human gaming tends to be more eclectic but is usually more experiential than analytic,[7] with uncertain implications.

Alternatives to Scenarios in Divergent Planning Exercises

At least two newer methods developed at RAND have been used frequently to complement scenario-based planning. These are *uncertainty-sensitive planning*, developed by one of the authors (Davis) and Paul Bracken in the late 1980s, toward the end of the Cold War, as a structured way of getting people to think about uncertainties and develop appropriately adaptive strategies (Davis, 1994a,b), and the related *assumption-based planning* (ABP), which was developed by James Dewar and others[8] as a creative but structured mechanism for uncovering the assumptions underlying baseline strategic plans and then suggesting hedges. ABP is admirably documented (Dewar, 2003). Both methods are summarized briefly in a recent review (Davis, Egner, and Kulick, 2005). These approaches seek to address uncertainties systematically. Uncertainty-sensitive planning seeks in particular to encourage planning for both branch points and unforeseen shocks. Assumption-based planning tries to uncover *all* salient assumptions, the implications of their failure, and possible signposts of impending failure. These methods are broad and somewhat rigorous, as contrasted to exploiting particulars and stories. Although they complement scenario-based planning and gaming, their effective use requires some of the same talents and attitudes, such as creativity and disdain for conventional wisdom and conventional tactics.

Exploratory Analysis in Search of Flexible, Adaptive, and Robust Strategies

As a next step in systematizing planning under uncertainty, RAND has also done a great deal of work on *capabilities-based planning* (CBP) (Davis, 1994a, 2002a), which was developed to move Defense Department planning beyond slavish adherence to standard scenarios. CBP was finally mandated in the 2001 *Quadrennial Defense Review* (Rumsfeld, 2001) and has, to a considerable degree, been implemented. A variant called *adaptive planning* now plays a central role in military operations planning. As developed at RAND, a key element of CBP analysis is conceiving the relevant "scenario space" with the intention of *exploratory analysis* to evaluate alternatives throughout that space.[9] This means considering not just one or two scenarios, but as many as necessary to cover the space. This idea is quite relevant to this report. Significantly, however, CBP's scenario spaces are typically oriented more to the parameters of military planning and analysis than to those of world futures. The methods used for CBP have been described as analogous to those used by a designer attempting to understand, for example, the operational envelopes of different future-aircraft candidates. CBP considers diverse crises and

---

[7] An exception is *foresight exercises*, which are intended to be analytic (Botterman, Cave, Kahan, and Robinson, 2004).

[8] The late Carl Builder contributed to ABP, building on his earlier discussion of scenarios' insidious effects and the need to better characterize their implications (Builder, 1983). He often emphasized, in criticizing scenario-based analysis, that "if you buy the scenario, you've bought the farm."

[9] This work on exploratory analysis was an outgrowth of research during the 1980s on the RAND Strategy Assessment System (RSAS), which emphasized *multiscenario analysis*.

conflicts and a richly differentiated set of operational circumstances but has not typically dealt with alternative economic, political, or social developments per se.

The central theme that has emerged in this work at RAND on CBP and uncertainty-sensitive planning is the need to emphasize strategies that are flexible, adaptive, and robust—i.e., FAR strategies—rather than strategies tuned to a particular expectation of the future. The adjectives here are meaningful and distinct, not clichés. *Flexible* relates to being able to use one's capabilities for different missions, including new ones; *adaptive* refers to the ability to accommodate readily to circumstances very different from those anticipated; and *robust* refers to being able to proceed effectively despite setbacks, such as a surprise attack or an unanticipated failure of some important system.

If a FAR strategy is represented as a plan, it may have a number of contingent decision points, with the plan proceeding along one or another branch thereafter, depending on circumstances at the time. A FAR strategy may not be optimal for the future in which it turns out to be operating, but it will be reasonably good—i.e., "good enough" in the sense of what Herbert Simon discussed as "satisficing" when introducing the concept of bounded rationality (Simon, 1982). Yet another aspect of a FAR strategy is that it includes hedges, perhaps in the form of reserve capabilities or in the form of just-in-case actions that make it possible to adapt quickly and reasonably well even to events that were not directly foreseen, occurring at times that were not anticipated.

This emphasis on FAR strategies constitutes a paradigm shift for those more accustomed to being provided official projections of the future, well-defined missions and tasks, and a few well-defined scenarios and authoritative databases with which to test alternative strategies or forces.

Exploratory Modeling

In a parallel stream of work in the 1990s, RAND also developed closely related methods for *exploratory modeling* (Bankes, 1993), with early applications primarily to subjects such as global warming, long-term forecasting, and business. The work also emphasized strategic adaptive planning, or what is sometimes called *robust adaptive planning* (RAP) (Lempert, Popper, and Bankes, 2003; Lempert, 2002). This emphasis is essentially equivalent to the emphasis on FAR strategies mentioned above.[10]

Although there have been substantial overlaps, exploratory modeling has sometimes been distinguished from exploratory analysis by research in which entirely different and competitive models have been compared in the same study. Two lessons from that work are that

- Strategies with simple built-in rules for adaptation are often superior to strategies based on a particular concept of the future.
- It is often possible to characterize meaningfully the conditions under which one strategy would be preferable to another in wagering terms, as in "Unless you believe that the chances of . . . are greater than one in ten, then . . . ."

---

[10] In the context of robust adaptive planning, *robust* means likely to work reasonably well in a wide range of futures, a somewhat different meaning from that in *flexible, adaptive,* and *robust* as we have defined those terms.

MSG for Strategic Planning: The Next Step?

It was against this background of research that we took on the challenge of developing methods and tools for MSG and the characterization of resulting scenario landscapes. The remainder of this report proceeds as follows. Chapter 2 extends the theory of exploratory analysis and the related activity of MSG. The focus is on the early, creative aspects of strategic planning, not the more-detailed aspects associated with follow-up construction of procedural plans. Chapters 3 and 4 describe two experiments with MSG undertaken to test the approach of Chapter 2 in very different domains. Finally, Chapter 5 draws some conclusions and presents some recommendations for future research.

2. A Preliminary Theory for Using Massive Scenario Generation

An Overall Process for Exploiting MSG

A Model to Create Scenarios

As indicated in Figure 2.1, we often see MSG as merely one part of a larger process of work in search of FAR strategies, although it serves other purposes as well, which we discuss later. Starting at the top left of Figure 2.1, having a model of the problem area is crucial,[1] although

Figure 2.1
MSG as Part of a Process for Finding FAR Strategies

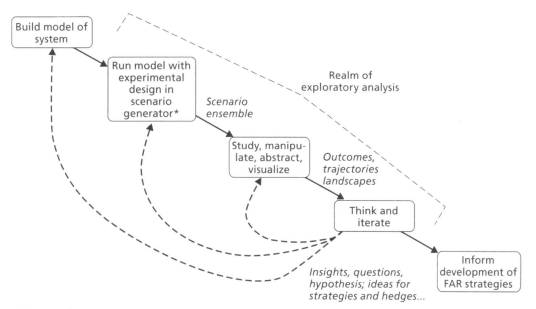

*The step of "massive scenario generation."
RAND *TR392-2.1*

---

[1] We decided on this approach only after considerable discussion. In principle, after all, knowledge can be accumulated without constructing a comprehensible causal model. One can imagine collecting innumerable fragments of knowledge and opinion from experts, operating on that knowledge base with computer programs, employing MSG, and coming up with insights. Something analogous was once at the core of artificial-intelligence expert-system concepts. Over time, however, people concluded that it was essential to have structure and explanation; merely having a machine serving as an answer generator was not satisfactory. The same point arguably applies in the current context.

the model may or may not exist at the outset of work. If it does not exist, work begins by conceiving the dimensions of the problem, drawing upon methods such as free thinking, brainstorming, gaming, and reading history or science fiction. The result is a first-cut model of the problem's system which is later iterated.

A Scenario Generator

After obtaining a model, the process calls for a computer device, a "scenario generator," to create an ensemble of scenarios sampling the space implied by the specification of an experimental design. The resulting scenarios are not simply a re-expression of inputs, because the specification may require considering all plausible combinations of certain variables, only some of which have previously been imagined explicitly. Further, allowance may be made for "wild-card" events.[2]

Figure 2.2 shows schematically that a scenario generator views the model as a mere black box: It specifies some inputs that are fed to the model, and the model runs and produces outputs; the scenario generator collects those input/output pairs and puts them in a database where they can be subject to post-processing, i.e., manipulated for the sake of visualization, reporting on particular issues as a function of scenario inputs, and so on. That post-processing can be interactive, with the analyst viewing standard reports and then deciding what special

Figure 2.2
Relationship Between Scenario Generator, Model, and Human
(process is iterative)

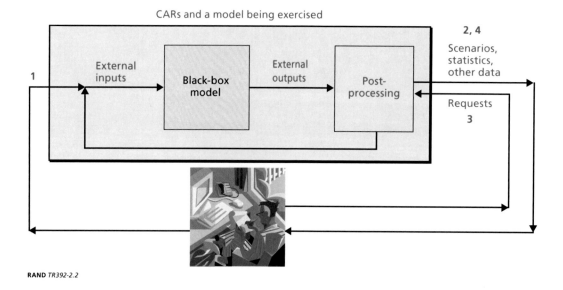

RAND *TR392-2.2*

---

[2] An analogy may be helpful: Given a network of paths from one village to another, a "specification" may be to generate all of the reasonably direct routes, e.g., all routes that never involve going backward for more than one mile. The result might include peculiar but interesting routes that involve temporary backtracking but overall efficiency or experiential virtues. And if the wild card of occasionally going outside the path network is permitted (e.g., random movement off-road to join a path segment not easily reachable otherwise), some routes discovered will not previously have been charted.

questions to ask or hypotheses to test. The analyst initiates the process in Figure 2.2 (step 1 in the figure) with an experimental design (i.e., a set of runs with systematically varied values of the input variables) or instructions on what kind of experimental design the scenario generator should construct. Later (step 2), the analyst looks at resulting data from the scenarios, makes new requests (step 3)—for either data analysis or scenarios—and then reviews the results (step 4). This continues as needed.

We used two scenario generators in this work: Analytica® and CARs®. Analytica[3] is a modeling system that includes scenario-generator and analysis capabilities. CARs[4] is an advanced research tool for scenario generation and analysis. It *uses* models developed separately, but its function is to drive those models systematically and assist in interpretation of resulting data. Analytica more than suffices for many applications of exploratory analysis, but in this study we were seeking to push toward the frontiers and needed the improved graphics, greater power, and better visualization and search capabilities provided by CARs. Those capabilities are best illustrated by the examples in Chapters 3 and 4.

Tools for Studying the Ensemble of Scenarios and for Recognizing Patterns

After generating the ensemble, the next step is to study it using analytical tools that manipulate and abstract[5] from the ensemble.[6] After thought and synthesis, one may obtain alternative, meaningful views of the "analytical landscape" painted by the scenario ensemble and analysis. For the intelligence community, the result may be insights that can help inform strategy, e.g., suggested hedges. In other domains, such as course-of-action development for policymakers or military commanders, development of FAR strategies may be part of the process itself.

This process is and should be highly iterative, as suggested by the feedback arrows in Figure 2.2. The model should be enriched over time, and the fruits of MSG should similarly increase.

Because all of this begins with a model, however, let us discuss in more detail what we mean by a model.

Approaches to Model-Building for MSG

Model Types

Modeling is central to our approach, but models come in many sizes and shapes. This is significant because, as noted above, models can either help or hurt the process of using MSG to generate insights.

---

[3] Analytica® is licensed by Lumina (www.lumina.com). It was originally developed as Demos at Carnegie-Mellon University by Max Henrion and Granger Morgan.

[4] CARs® was developed by Evolving Logic (www.evolvinglogic.com), of which one of the authors (Bankes) is a principal.

[5] To "abstract" is to simplify, while capturing the essence. This may be accomplished, for example, by averaging over some variables, ignoring fluctuations or reducing resolution (Sisti and Farr, 2005).

[6] Both Analytica and CARs include such tools. Other tools that can be brought to bear include standard statistical programs such as STATA® or, e.g., the familiar Microsoft Excel®.

Figure 2.3 indicates schematically how we grouped model types in thinking about MSG. It distinguishes to first order between causal and noncausal models. Many other typologies are possible (e.g., quantitative and qualitative), but it is the causal versus noncausal that we wish to highlight here.

Causal Models

Consider first (top of Figure 2.3) causal modeling. It may do a good job of itemizing the system's objects, their attributes, and the processes that change them. The model, however, assumes precise knowledge of all or nearly all of those ingredients. This is not helpful for planning under uncertainty. Moving downward in the figure, we may study a problem with a causal model and the method of exploratory analysis (sometimes referred to as "sensitivity analysis on steroids"). Uncertainties are confronted directly, and a vast range of possibilities is revealed. Even this, however, is not enough if the nature of the system itself is uncertain. Many real-world systems are not stable. Instead, new objects (e.g., leaders) appear and disappear; new processes come into being, in part as the result of technology and innovative thinking. Moreover, some of what happens is due to hidden variables or chance. The paradigm of complex adaptive systems (CAS) is notable primarily for having emphasized this reality.[7]

This said, models—even CAS models—can either help or limit the value of MSG. The concept of system modeling is now widely understood, and lip service is paid to construing the system broadly, but representing the possibility of fundamental changes in system *structure* and *processes* has not been much discussed. Such matters remain at the frontier. In our view, there is much that modelers still need to learn and assimilate from the world of human gaming, where innovations and boundary-breaking are routine. Overall, we urge a man-machine approach to inquiry, rather than something based more exclusively on computation. Consistent with this, we show a category of system modeling that proceeds with more humility by attempting to represent various aspects of both static and dynamic structural uncertainty.

Figure 2.3
Different Types of Model

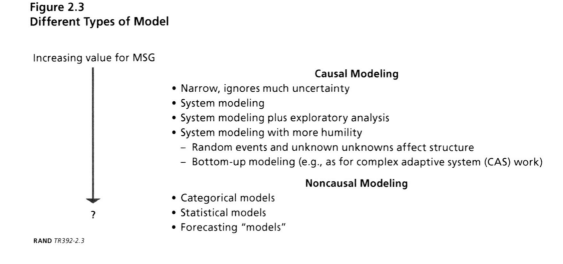

---

[7] Many references on CAS now exist; Holland and Mimnaugh (1996) remains an excellent introduction.

To fully explore the possibility space in problems of interest to policymakers, it will surely be necessary to use the paradigm of CAS and related models.[8] Generating massive numbers of scenarios based on structurally rigid models will not suffice in the long run.

We are confident that the value of MSG will increase as the richness of causal models is increased to include both parametric and structural uncertainty (moving downward with the arrow in Figure 2.3). This said, there are other, noncausal approaches to modeling (bottom of the figure), which we did not specifically examine. We are uncertain but skeptical about the value of MSG based on such models.

Noncausal Models

"Categorical models" (e.g., biological taxonomies) describe systems in terms of dimensions, factors, or attributes. The protocol used by an answering service to specify action in response to different types of telephone call is one type of categorical model. In political science, "structural realists" view the world through attributes of a situation, such as geography, resources, and power.

Statistical models summarize empirical or subjective data. In medicine, a statistical model might show an association or correlation between future heart disease and high cholesterol, a family history of heart attacks, and a sedentary lifestyle. In military affairs, a statistical model might suggest that occupying a country such as Iraq would require a large army—not necessarily for warfighting, but for stabilization and recovery (Dobbins et al., 2003). In intelligence, such a model might suggest a strong correlation between various measurable indicators and the likelihood of insurgency or revolution in a third-world country (de Mesquita, 1983). One could use statistical models to generate what might be called scenarios.

Forecasters and futurologists have also used the term *scenario* without explicit causal models. They may consider a few streams of development, such as demographics and technology, note that breakthroughs occur from time to time in technology and that major events such as economic depression or periods of boom sometimes occur, and construct possible futures in which the various streams unfold in different ways. The late Isaac Asimov exploited such thinking in some of his novels, as did Herman Kahn in his work (Kahn, 1983).

How Much Is Enough in MSG?

With this background on initial divergent thinking and modeling, the next question is, How does one specify the possibility space to be explored with MSG? Other questions include, How much is enough? and, How "massive" is massive enough? It is reasonable to assume that

---

[8] Agent-based models (ABMs) often have dubious usefulness as predictive models, but they are quite good as scenario generators because they allow insertion of both knowledge and speculation about agent motivations and capabilities. In some scenarios, they can be the engine for nonintuitive emergent phenomena. To use the terminology of the current report, they introduce changes of model *structure* that would not ordinarily be considered. The relevant literature is quite extensive, covering, e.g., artificial societies (Epstein and Axtell, 1996), social networks (Prietula, Carley, and Glasser, 1998), counterterrorism (MacKerrow, 2003), military problems (Ilachinski, 2004), and, from RAND work, technology diffusion (Robalino and Lempert, 2000) and the emergence of near-peer competitors (Szayna, 2001). Some reviews also exist (e.g., Uhrmacher, Fishwick, and Zeigler, 2001; Uhrmacher and Swartout, 2003).

increasing the richness and resolution of the scenario space will add potential value, but when will a state of diminishing returns be reached—especially when we take into account the need to comprehend and explain the results?

In response to the question, How much is enough? our initial hypothesis was that the potential value (i.e., the information content) of MSG increases with dimensionality and resolution roughly as follows: We suspect that there is likely to be substantial value in adding additional categories and distinctions (dimensions)—certainly in going from 2 to 10—but that the relationship is logarithmic. We doubt that 10,000 dimensions will be materially better than 100.

We also expect that increasing granularity or resolution along any given dimension will not be particularly worthwhile—unless, for some reason, the landscape is rough in spots, with the output under study varying rapidly with the variables being considered. Thus, when exploring a given uncertainty, we expect that looking at endpoint values, or endpoint values and a few intermediate points, will ordinarily be adequate for exploration. For example, sample warning times for a war might have values of zero, seven days, one month, and one year, but not values of 6.32 days or 1.46 months. As a counterexample, the behavior of a crowd might vary with small changes of circumstance, going from docile to exuberant to rowdy in seemingly erratic ways.

Both hypotheses are reflected in (a) of Figure 2.4. We also hypothesized the relationship shown schematically in (b) of Figure 2.4. The idea is that practical value—i.e., value in

Figure 2.4
How Much Is Enough, and Even Too Much?

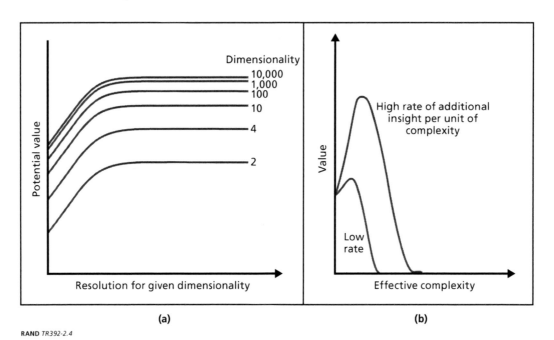

(a)   (b)

understanding—increases as we increase effective complexity from levels of triviality but then decreases as the complexity overwhelms us: If we can't understand, then the potential value is not realized. The figure also suggests that there are better and worse ways for increasing complexity as we build models and generate scenarios. The issue becomes, qualitatively, the rate of additional insight per unit of complexity.

Methods for Making Sense of Complexity

The question of how to maximize insight per unit of additional complexity then arises. This bears on how we build models. The history of theoretical advances, which can be seen as advances in modeling, has much to tell us. These advances include the introduction of vectors, matrices, tensors, and operators in mathematics and modular decomposition and system engineering in engineering.

More generally, abstractions play a key role when we attempt to make sense of complexity and the need exists to limit the number of "cognitive chunks" (Simon, 1981). This is a principal motivation for *multiresolution modeling* (MRM), which allows the user of a model to input variables at different levels of resolution: Although detail is sometimes necessary, comprehending issues is best done with a more abstracted view of the problem (Davis and Bigelow, 1998).

As a completely different kind of abstraction, and drawing from both psychology and everyday life, we know about the importance of using stories to convey concepts that transcend the stories themselves. We also know about the importance of visualization, experiential learning (as in war-gaming), and puzzle-solving.

With these points in mind, we chose a small number of techniques to use in sense-making within our short-duration experiment.

Four Methods

The problem here is, Given data for an ensemble of scenarios, how do we make sense of the results?[9] Any number of methods exist; the question is as open-ended as, How does one do analysis? but in this study we used four primary methods. These range from simple linear sensitivity analysis (often called *importance analysis* by those who include related mechanisms in software), to a variant in which one uses combinations of raw input variables, to advanced data analysis using filters, and finally to motivated metamodeling. The first three methods are in the nature of observational empirical data analysis;[10] the last is more like empirical analysis by a physical scientist, with significant but imperfect hypotheses about underlying phenomenology:

1. *Linear sensitivity analysis* as one measure of how "important" different variables are in determining outcomes.

---

[9] Similar questions are important in command-and-control research (Leedom, 2001).

[10] Other methods that are sometimes used to discover ways to reduce dimensionality of problems include factor analysis and multidimensional scaling. Short descriptions exist on the Web (Darlington, 2004), and a number of books on such methods are available (e.g., Morrison, 1990).

2. *Extended linear sensitivity analysis*, using "aggregation fragments" hypothesized to be better predictors than raw input variables.
3. *Advanced data analysis with filters*, using methods adapted from those in data-mining and cluster analysis, both to help identify important variables and to find multidimensional patterns.
4. *Motivated metamodeling.*

Linear Sensitivity Analysis

If a model is thought of as a function $y = F(x)$, where x is the vector of all input variables and y is the vector of all output variables, either at the latest point in time or along the way, then one can define a measure of the sensitivity of a particular outcome variable y_i to the input x_j as

$$S_{ijk}^0 = \left(\frac{\partial y_i / y_i}{\partial x_j} \right)_{x',k} dx_j$$

That is, the sensitivity measures the fractional change in y_i caused by a small fractional change in x_j. This measure, however, depends on the values of all of the other input variables, i.e., on x'. It also depends on the particular scenario, k, in the ensemble of cases generated. In a multidimensional problem, such a complex measure of sensitivity might be too complex to be useful, so it is customary to define instead a sensitivity S that is the average over the other variables. If there are N scenarios, then

$$S_{ij}^{(1)}(x_j) = \frac{1}{N} \sum_k < S_{ijk}^0 >$$

where $< >$ denotes taking the average over the other variables. Since the result still depends on x_j, a final measure averages over that:

$$S_{ij} = < S_{ij}^{(1)}(x_j) >$$

Yet another approach involves fitting a linear regression to the "data" on the function's behavior. The resulting coefficients are the measures of the sensitivity to each of the input variables.

Such measures of sensitivity are useful but are often difficult to interpret and may even be misleading. If a variable x_j is identified as important by linear sensitivity techniques, it very likely is. However, if a variable is not identified as important, that may be an error. For example, if a variable has very large positive and negative influences, depending on other factors, the apparent significance of the variable may be small because the positives and negatives cancel out in the process of averaging.[11]

As another example, Figure 2.5 shows a graph of $z = x^2 y^{1/2}$. The origin is in the lower left corner. The x axis goes into the sheet, the y axis comes out of the sheet, and the z axis is vertical. The sensitivity of z to x is substantial for small values of y. However, for larger values of y, the sensitivity is much less. Suppose, now, that we asked for the sensitivity of z to x. If we averaged the value of the slope of z(x) over the range of y values shown, we would get an average sensitivity only about half as great as that for y = 0. If, moreover, we averaged over values of x, we would get a value much less than that for x = 0.

Using Aggregation Fragments

A limitation of out-of-the-box linear-regression methods is that they encourage building statistical models of a phenomenon based on a linear sum of the independent variables or, in more-sophisticated treatments, by polynomial sums. That is, if the inputs are x_1, x_2, and x_3, one might try a linear sum of these or a more complicated sum that included x_1^2, $x_1 x_2$, etc.

Figure 2.5
Graphic Illustration of Problems in Averages

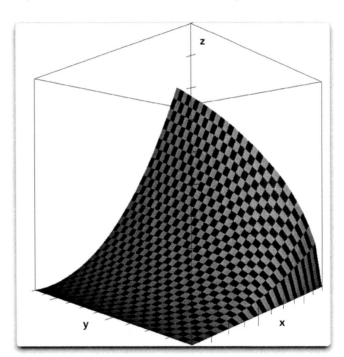

RAND *TR392-2.5*

---

[11] Another problem is correlation. Although inputs to a function may be mathematically independent, the real-world probability of them having specific values may be highly correlated. For example, in security problems, it may be wise to assume that an adversary would be able to arrange circumstances for an attack so as to minimize warning time *and* involve a good deal of confusion, delaying decision, *and* involve deception, causing initial reactions to be inappropriate.

If enough data exist, regression methods can find the coefficients that will cause a best fit of such an expression to the data. The result may or may not have any physical or explanatory value. A first improvement can be made by identifying natural *combinations* of the independent variables that seem likely to matter. In the physical sciences, for example, a phenomenon may vary with a speed times a time, divided by a natural length of the system, i.e., by (V)(t)/L. In biology, an important fragment might be the length of time a chemical must be in the body to have a therapeutic effect divided by the time it takes for it to be removed by natural processes of the stomach, intestines, liver, and kidneys. In any given problem domain, experts will often be able to make good guesses about such fragments. If so, then the fragments are good candidates to be used in data analysis by generalized linear regression in which one of the variables tested might be, say, x_1/x_2 or $x_1 x_2 x_3/(1+x_4)$—whatever fragments make sense.

Using Advanced Filters

A third way to find important variables is to apply various filters. The most familiar example might be a noise filter. When listening to a very noisy telephone conversation, one may not be able to recognize voices or understand what is being said. After applying a noise filter, however, both voices and words may become understandable. Many other filters exist or are possible. In digital manipulation of imagery, for example, applying a filter may sharpen boundaries and other linear features. Applying another filter, one that increases contrast, may cause previously merged areas to be seen as distinguishable.

An interesting type of filter that we used in our experiment comes from the worlds of data-mining and cluster analysis. Suppose one has both input and output data for some system and is unable to find much structure with usual methods. The problem may be that there are different clusters within the data, for each of which separately one could readily find structure. That is, the data in each separate output cluster might be a relatively simple function of the inputs. The Patient Rule Induction Method (PRIM) (Friedman and Fisher, 1999) can be applied to a data set to find sets of input-value ranges that generate a particular range of output data, say, high values. PRIM essentially finds the n-dimensional box of input-variable space in which most cases of the particular output range are to be found and in which most outcomes indeed exist.

This type of search is imperfect: Topologies exist for which it will give misleading results (e.g., topologies containing a toroidal structure, or donut). Also, it is only one of many useful algorithms. Nonetheless, it has proven useful in exploratory analysis, in significant part because its findings are readily explainable (Lempert, Groves, Popper, and Bankes, 2006), as in, "To get a good outcome ($y > y_{min}$), x_1 can have almost any value, but x_2 almost always must be within a range x_{20} to x_{21}, and so also x_3 must almost always be in the range x_{30} to x_{31}."[12]

Motivated Metamodeling

The methods discussed in the preceding sections are all useful in finding important variables in terms of which to interpret data. In some cases, using the results can yield graphical depictions

---

[12] Other methods might, for example, search for rules using "or" expressions, such as "High values of the output Y are associated with values of X within this range *or* that range."

that show strong patterns and, in some respects, make sense of the data. However, if we seek to understand the results at a deeper level, what we actually seek is a simple theory, not just some visual patterns. Arguably, such a theory would take the form of either a relatively simple and understandable mathematical equation or a relatively simple computer model. In either case, we would hope for a theory based heavily on a relatively small number of variables—something that might be called, variously, a low-resolution model, a reduced-form model, an abstract model, or a metamodel. For the remainder of this subsection we use the term *metamodel*.

Metamodels are models of models. If one has a complex model with hundreds of input variables, the model's outputs may in fact be described well by a relatively simple function of only a few variables. That is, the full model may be incomprehensible, but its "behavior" may be understood through that simpler function (a metamodel). The function (often called a *response surface*) is usually found by applying standard statistical methods such as multivariate regression to a sample of output data generated by the original model. Thus, the function is a regression equation and will often not have any straightforward phenomenological interpretation. If this is the case, the metamodel may have limited value: It tells no "story," it provides no meaningful "explanation," and it has no face-validity.

Motivated metamodeling is the name that has been given to an alternative approach to the same situation (Davis and Bigelow, 2003). In this approach, one draws upon an understanding of the problem domain to hypothesize a reasonable, albeit approximate, analytical structure. Instead of attempting to fit a linear or polynomial regression of inputs, $x = x_1, x_2, \ldots$, one might hypothesize the form of the sought-for simple function that explains the outcomes reasonably well. An example might be $e^{-C_1(t/\tau)} \cos(c_2 \omega t)$, the only point of the example being to illustrate structures (a product of exponential and trigonometric functions) that might be justified by knowledge of the problem. Presumably, such a hypothesized form would be based on a simplified theory, in which case—if the structure proves accurate—one would have a way to explain results. This is a new name for an old method, one often used by scientists. Some observed properties of relatively low-density gases, for example, can be described as the behavior to be expected from the ideal gas law and some minor corrections proportional to the density of the gas. The explanation, then, is the explanation of the ideal gas law itself, plus the observation that corrections reflect collisions among the molecules, the frequency of which should be proportional to the density of the gas (at relatively low densities).

Ideally, motivated metamodeling should build into the hypothesized form of the metamodel various correction factors and coefficients so that statistical testing can reveal shortcomings of that form, not merely find "fudge factors" to make it work fairly well. In the current work, we used a metamodel form as follows: If z is a vector of data to be understood in terms of a vector of independent variables x, and if F(x) is the functional form motivated by an approximate theory, then we would first try a motivated metamodel, perhaps as follows:

$$z = C_0 + MF(x)\left\{1 + \sum_i C_i x_i\right\}$$

If the best fit to the data came from a value of M = 1 and zero values of the C coefficients, then F(x) would be perfect. However, if these statistically derived coefficients came out otherwise, then F(x) would be seen as an approximation. For example, the result might be a good fit with M = 2, C_0 = 0.25, C_4 = 0.2, and the other coefficients being very small. In that case, one might say that the motivated metamodel is rather accurate except for a bias error corrected by the C_0 term and some modest additional dependence on x_4 that is not captured within the function F(x). Examples are given in later sections.

Dual-Track Experimentation

We conducted two parallel experiments in massive scenario generation (MSG). The first experiment began with a partially developed model of terrorism and counterterrorism intended for use in the policy community. It proved necessary to add features to that model in order to better represent uncertainties, especially structural uncertainties that would not usually be considered.

The second experiment began with writing a report on conditions of the next nuclear use (NNU) based on an unpublished paper on the same subject. We saw this as akin in spirit to beginning not with a well-defined analytical structure for the system, but rather with only partially structured insights from a workshop report or war game. As written, the report had no causal structure and could be seen as describing a categorical model.

Attributes of the two approaches are given in Table 2.1.

Figure 2.6 seeks to motivate the approach. After considerable thought, we concluded that the virtues of a given approach to planning under uncertainty can be characterized by four metrics, the axes in the figure. There is value in an approach that does not require having a preexisting model. In crises, for example, it is often the case that no adequate model of the issues exists (or, at least, no model that is well known and understood by those involved). It is also desirable that the activity consider many dimensions of uncertainty and that the scenario space or possibility space implied by those dimensions be explored systematically. At the end of the day, it matters also whether the quality of final knowledge is high.

Table 2.1
A Two-Track Approach to Experimentation

Starting Point	Attributes
1. Analytical counterterrorism model with extreme parameterization	• Base version is like common models, with a particular image of reality • Expanded version can explore under deeper uncertainty
2. Essay-style report on circumstances of NNU	• Akin to beginning with the report from a strategic-planning exercise

Figure 2.6
Contrasting Virtues of Two Approaches

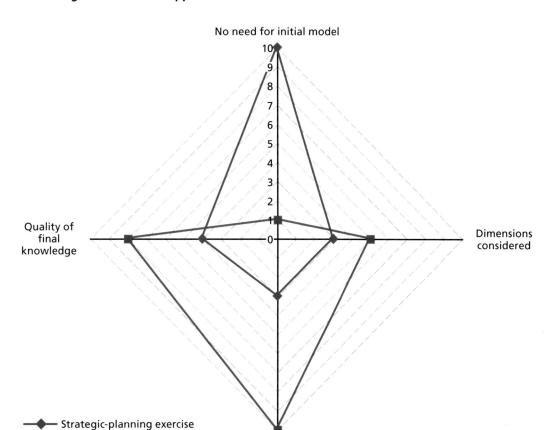

RAND *TR392-2.6*

Figure 2.6 shows the four metrics as axes of a radar chart. Goodness corresponds to moving outward on those axes. The chart is illustrated with a contour for a strategic-planning exercise. The assertions here are that strategic-planning exercises based on a few scenario dimensions (the blue contour in the figure) do not require an initial model, explore only a few dimensions of uncertainty, do not explore the space of possibilities significantly, and result in a level of knowledge that is significant but far short of what one might like. The knowledge gained may, for example, implicitly overweight the particular scenario(s) examined in the exercise. Most seriously, it may not even recognize the existence of many key factors.

One can do better, e.g., by bringing in aspects of human gaming and considering more dimensions than are used as the formal basis for discussion. Sometimes, the resulting knowledge may be quite good for the purposes intended by the exercise. This said, Figure 2.6 is probably fair in suggesting the relative strengths and weaknesses of a typical strategic-planning exercise based on only a few dimensions of the scenario.

The red contour shows the strengths and benefits of an approach that uses modeling from the outset. Needing a model is a disadvantage, but the model can consider more dimensions than are normally considered in scenario-based strategic-planning exercises. The model can be used to explore quite a large space, and at the end of the day, the quality of final knowledge may be rather high.

The idea, then, was that our two experiments would give us experience with both classes of activity.

Where Is the Value in MSG?

To conclude this chapter, let us now turn to the question of what constitutes "value" in MSG. Do we understand this? Could we measure it, at least in principle?

In the terms of Figure 2.6, our expectation was that MSG with a "good model" would be a very useful extension of what we have described elsewhere as exploratory analysis (or as exploratory modeling). We saw the primary payoffs as being in the realm of an improved capability to cover the parameter space and improved analysis of "data" from exploration. We saw the payoff of MSG, when starting from a poorer base, such as a strategic-planning exercise, as much more qualitative in nature: MSG should be able to increase dimensionality and exploration substantially, and the quality of final knowledge to at least some degree. How much was not clear.

Table 2.2 suggests thinking about the value of MSG and the subsequent synthesis in three categories—intellectual, pragmatic, and experiential—only some of which we had in mind prior to the project.

Intellectually, we can hope that MSG would increase our understanding of system phenomena, which in turn would suggest what questions to ask, what tests to conduct, and what developments to monitor.

Table 2.2
Measures of Value for MSG

Intellectual
Understanding phenomena Knowing what to ask for, test, and monitor
Pragmatic
Enriching intuitive pattern libraries to improve at-time tailoring and adaptation Enriching strategies with branches and hedges Narrowing crucial issues to test or explore in games or experiments Designing mission rehearsals
Experiential
Designing richer training, education, research, and socialization exercises — Widening sense of the possible — Practicing problem-solving for plausible but nonstandard cases

Intellectually, we can hope that MSG would increase our understanding of system phenomena, which in turn would suggest what questions to ask, what tests to conduct, and what developments to monitor.

In the pragmatic category, we anticipate that MSG could be a good mechanism for increasing the extent of vicarious experience and thereby increasing the library of patterns that decisionmakers and analysts have in their minds. These patterns help them "connect the dots" and make sense of what might at first seem chaotic. The empirical literature documents the significance of this ability to expert performance (Klein et al., 1993). As indicated at the beginning of this report, we also have in mind that MSG could help enrich strategies by suggesting unappreciated possibilities and causing strategies to have additional branches and hedges. Even the most action-oriented decisionmakers are often willing to prepare adaptations that might be necessary to achieve their preferred core strategy, as long as they do not perceive the hedges as undercutting the strategy. Related to this is the process of identifying the crucial issues to be tested or explored in games, experiments, or actual operations. Finally, mission rehearsals can sometimes be improved by exposing participants to a stressful set of intramission challenges that may be encountered.

This overlaps with the experiential category of value. We see MSG as being a useful mechanism for enriching training, education, and research and development (R&D) by opening minds and giving participants hands-on experience in solving unanticipated problems, and doing so together with people they may need to work with in real-world crises. Historically, such skill-building and team-building have been valuable in preparing for crisis operations, either in business or in government.

3. Experiment One: Exploratory Analysis with an Epidemiological Model of Islamist Extremism

A Model of Terrorism

Our first experiment involved an epidemiological model of terrorism and counterterrorism, one begun under RAND's Internal Research and Development (IR&D) for the Department of Defense.[1] The model was substantially extended and enhanced for the present study.

The basic idea of the model was to discuss Islamist extremism by analogy with disease.[2] The disease emerges; it may or may not spread, and it may even take off. Even if it does take off, it may reach a peak and then diminish over time. Policy measures may or may not affect all of this. In medicine, the policy measures include inoculation, quarantine, treatment, and public-health sensitization campaigns. When dealing with extremism, the policy measures may involve counterideology campaigns; reducing contact between extremists and others, whether in prisons, from the pulpit, or through the media (quarantine); engagement (treatment?); and infiltrating organizations (arguably somewhat analogous to a sensitization campaign).

Motivated by the desire to maximize the portion of possibility space addressed, we added numerous features to reflect deep uncertainty about the various parameters. We also represented the possibility of significant "exogenous effects" that, in the real world, can make a big difference to the time history of extremism as a disease but that cannot realistically be predicted. Examples of such events might be the United States intervening in another Muslim country, a surprise agreement emerging between Palestine's Hamas party and Israel, or a special individual rising to leadership in the Muslim world.

The model may be applied to an expatriate population, such as Muslims in France, or to the population of a country such as Egypt or Turkey. A given person in the population of interest is considered to be in one of eight states, as shown in Table 3.1. The person might be in an ideological state referred to in brief as oppositional, passive, sympathetic, or extremist. These terms relate to degrees of acceptance of militant Islamism. If extremism is regarded as the disease, then a sympathizer has the disease but is not yet in the most advanced stage. A sympathizer might harbor extremists, including terrorists, but would not himself engage in extremist activities. Extremists actively pursue the cause and may even participate directly in

---

[1] Paul K. Davis, "An Epidemiological Model of Terrorism and Counterterrorism," unpublished, 2006.

[2] This approach was motivated in part by earlier work with simple analytical models (Castillo-Chavez and Song, 2003).

Table 3.1
States of Ideology and Immunity

Ideological state	Immunity state
Oppositional	Susceptible
	Immune
Passive	Susceptible
	Immune
Sympathetic	Susceptible
	Immune
Extremist	Susceptible
	Immune

terrorism. Someone who is passive is not infected and tries to avoid taking sides. One who is oppositional actively works against the disease of extremism, perhaps by informing authorities about dangerous activities.

An individual in any of the four ideological states may be either susceptible to further infection or immune. In practice, the state of an immune extremist makes little sense and is treated as empty.

Figure 3.1 is a screen shot of the model's user interface. The model was written in the Analytica modeling system, which is particularly well suited to building models for exploratory analysis. As the figure indicates, the primary input parameters are the initial population fractions, the contagion rate, the "natural" recovery and immunization rates, the effects of various abstract policies on the contagion and recovery rates, and the probabilities and consequences of random exogenous effects. The latter are modeled by assuming that in any given year, a "good" or "bad" event, or both, may occur. A good event decreases contagion rate and increases recovery rate; a bad event works in the opposite direction.

The model has additional parameters within the Tuning Parameter module, but these 18 basic parameters (the Initial Population by Ideological States has three independent parameters) were more than enough for our experiment.

The model's outputs include trajectories of the population fractions and various specialized measures, a few of which are indicated on the right side of the interface.

Exploration is accomplished by establishing a set of values for each parameter and having Analytica run cases for each of the combinations. If all of the 18 parameters were given two values, the result would be 2^{18} runs (about 300,000). In practice, we did most of our work with Analytica holding some of the parameters constant and running 1,000 or so cases. With the advanced scenario-generator tool CARs, we routinely ran tens of thousands of cases (and used a Latin Hypercube design to improve the efficiency with which we sampled the space of cases).

Figure 3.2 shows a top-level view of the actual model. It is a type of influence diagram[3] showing graphically that the model is structured along the lines described. Each of the

---

[3] In the current work, our influence diagrams are reasonably close in spirit to those in system dynamics (Forrester, 1969; Sterman, 2000). In other applications, we use such diagrams more like what are sometimes called *cognitive maps* or like decision models (Davis, 2002b). In still other applications, such diagrams are used to help define Bayesian nets (Pate-Cornell

Figure 3.1
Model Interface: Inputs and Outputs

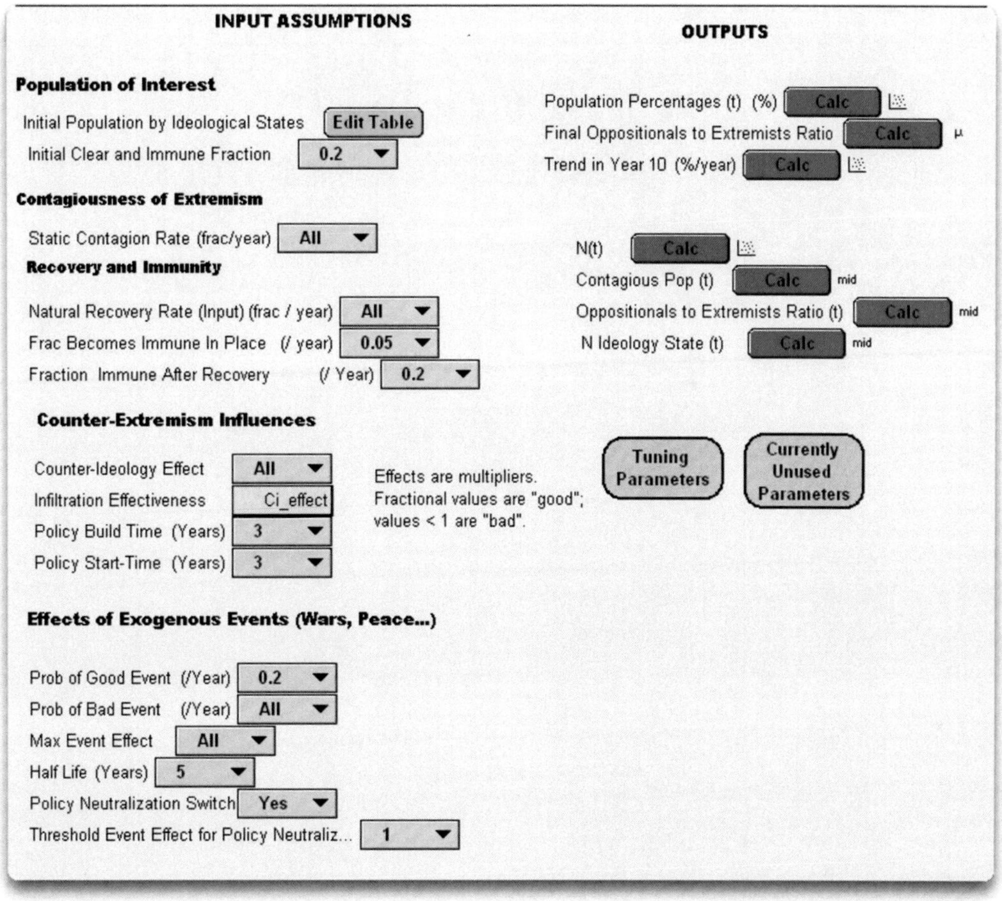

```
                    INPUT ASSUMPTIONS                              OUTPUTS

Population of Interest
                                                      Population Percentages (t)  (%)  [ Calc ]  📊
Initial Population by Ideological States  [ Edit Table ]   Final Oppositionals to Extremists Ratio  [ Calc ]  μ
Initial Clear and Immune Fraction    [ 0.2  ▼ ]       Trend in Year 10  (%/year)  [ Calc ]  📊

Contagiousness of Extremism
                                                       N(t)    [ Calc ]  📊
Static Contagion Rate (frac/year)  [ All  ▼ ]
Recovery and Immunity                                  Contagious Pop (t)   [ Calc ]  mid

Natural Recovery Rate (Input) (frac / year)  [ All  ▼ ]   Oppositionals to Extremists Ratio (t)  [ Calc ]  mid
Frac Becomes Immune In Place  (/ year)  [ 0.05  ▼ ]       N Ideology State (t)   [ Calc ]  mid
Fraction  Immune After Recovery    (/ Year)  [ 0.2  ▼ ]

    Counter-Extremism Influences

Counter-Ideology Effect    [ All  ▼ ]      Effects are multipliers.     ( Tuning       ( Currently
Infiltration Effectiveness  [ Ci_effect ]   Fractional values are "good";  Parameters )   Unused
Policy Build Time  (Years)  [ 3  ▼ ]        values < 1 are "bad".                       Parameters )
Policy Start-Time  (Years)  [ 3  ▼ ]

  Effects of Exogenous Events (Wars, Peace...)

Prob of Good Event  (/Year)  [ 0.2  ▼ ]
Prob of Bad Event    (/Year)  [ All  ▼ ]
Max Event Effect    [ All  ▼ ]
Half Life (Years)  [ 5  ▼ ]
Policy Neutralization Switch  [ Yes  ▼ ]
Threshold Event Effect for Policy Neutraliz...  [ 1  ▼ ]
```

RAND *TR392-3.1*

bubbles in the diagram is a module. By double-clicking on such a bubble, one can see the next level of detail.

Figure 3.3 is a screen shot from an illustrative use of the Analytica model. It shows the population of the various ideological states versus time, assuming an initial population of 10,000 people. The curve for oppositionals is seen to rise, leveling off at about 25 percent. The number of extremists (bottom curve) rises initially but then diminishes. The results shown here, however, depend also on the values of the various parameters shown at the top of the figure. By clicking on any of the "rotation boxes" there, one can change the value of that parameter, and the curves shift correspondingly. Because all cases have been run already, the data for these curves—covering thousands of cases—are in the computer's storage, and the curves change more or less instantaneously as one explores by clicking. An analyst can explore

and Guikema, 2002) or influence nets (Rosen and Smith, n.d.; Wagenhals, Shin, and Levis, 1998). We found all of these other methods to be powerful, but we have not used them here.

Figure 3.2
Top-Level Influence Diagram

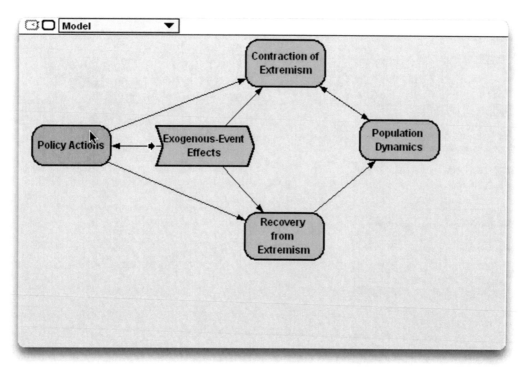

RAND *TR392-3.2*

the space of outcomes by closing the office door and concentrating for perhaps a half-hour. This type of exploratory analysis has now been used extensively over the past five to ten years (Davis, Bigelow, and McEver, 2001; Davis, 2002a).[4]

Figure 3.4 shows an aspect of the problem not highlighted in Figure 3.3, the huge variation in disease histories from one run to another. This figure shows the time dependence of the number of extremists for a set of five runs out of the thousands (remember that exogenous events occur randomly). We see that outcomes are drastically different—by and large, they tend to settle as very bad, intermediate, and very good across large numbers of samples.

---

[4] Although not much of an issue for the current epidemiological model, a very important issue more generally is mathematical convexity. If the outcome focused upon never decreases with increasing values of a parameter (i.e., is monotonic), then the relevant function is "convex," which simplifies exploratory analysis. However, in many problems, there are nonmonotonicities (nonconvexities). For example, decentralization in military command and control or in the execution of a counterterrorism policy might pay substantial benefits for a while (as noted by advocates of "power-to-the-edge" concepts of network-centric operations), but at some point, further decentralization could lead to local actions with strategically counterproductive consequences. That is, more of a "good thing" would in fact be bad. It is also possible for effects to oscillate rather unpredictably as a policy-related parameter is increased. This can happen, for example, due to the interaction of local decisions with knife-edge rules of engagement. Related phenomena have been studied in military command and control (Dewar, Gillogly, and Juncosa, 1991; Tolk, 2001). For nonconvex problems, the experimental design for generating cases must be considerably more complex and must involve larger sets of cases. Techniques such as Latin Hypercube design become important in mitigating growth in these cases. Such designs were used in the work with the CARs system discussed later in this report.

Figure 3.3
Populations Versus Time from One Scenario

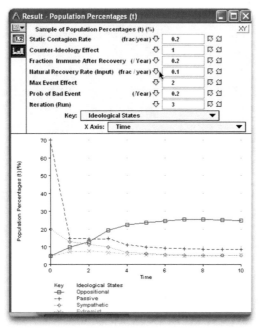

RAND *TR392-3.3*

The significance of this is that great caution is warranted in ascribing significance to "average" results. Nonetheless, we will do that to some extent for simplicity in some of what follows.

Making Sense of the Data from MSG

Initial Results

In what follows, we illustrate more-advanced methods of examining, manipulating, and abstracting from exploratory analysis. We simplify by focusing on a single measure of outcome, the ratio of oppositionals to extremists in the tenth year. Obviously, large numbers are good. Although this focus is a simplification, it is usually representative. We also move to use of the CARs platform because of its better graphics, ability to deal with massive numbers of cases, and extensive filtering capability. As noted earlier, however, CARs is not a model. Rather, it uses a model, in this case the Analytica model described above. In other applications, the model might be expressed in Excel, C, or some other language.

Figure 3.5 illustrates one of the initial outputs from CARs for the epidemiological model of extremism. It is a "dot plot" in which each dot represents the outcome of a single scenario. Unlike traditional two-dimensional outcome charts, this one uses color to indicate outcome, with green being good and red being bad. Both the x and y axes represent input parameters.

Figure 3.4
Run-to-Run Variation in Scenario Trajectories:
Prediction Is Clearly Inappropriate

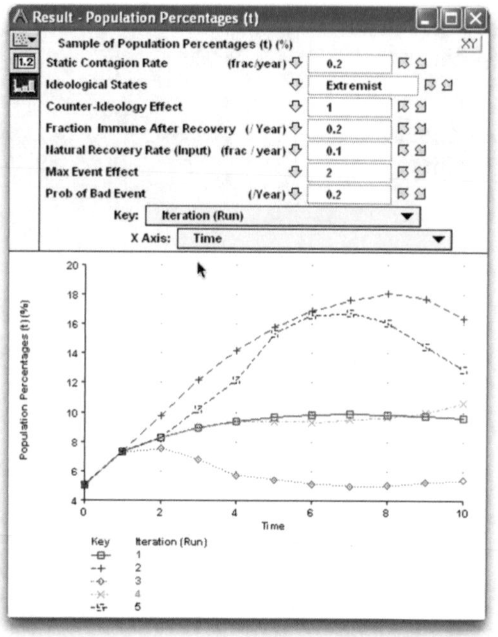

RAND *TR392-3.4*

In this case, we looked for outcomes when plotting the initial clear-immune fractions (y axis) versus the magnitude of policy effects (x axis).

The chart is very cluttered; throughout, one sees a mix of different colors (different outcomes). Basically, the chart is not useful. We show it merely to make the point that making sense of raw scenario data requires some effort.

The reason for the "noise" in this case is a matter of projection. The outcome (color) depends on the values of many variables. If, however, the points and colors are plotted in a two-dimensional display for an arbitrary set of axes, we see the projections of those points onto a two-dimensional plane of the n-dimensional volume. As a result, we can no longer see how the color of the points varies with the "other" variables. As an analog in a mere three dimensions, consider Figure 3.6. On the frontmost image, we see many green points on the left side (y) and many red points on the right (x); in the center image, we see the opposite. The projection onto a single two-dimensional xy plane combines these to create the rearmost image, in which we see a mix of red and green points throughout. That result looks more random than the original because of z, the "hidden variable."

This illustrates in a trivial case what we must deal with more generally in trying to make sense of MSG "data." Unless we visualize the data along the right dimensions, we do not see patterns. The right dimensions can be found using statistical methods, intuition, theory, or a combination of those and experimentation.

Figure 3.5
A First Dot Plot: No Obvious Pattern Is Discernible

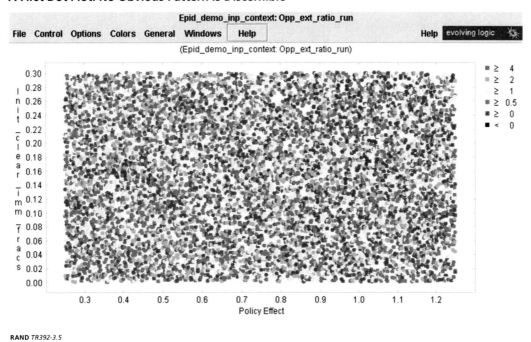

Figure 3.6
Effects of Projection and Hidden Variables

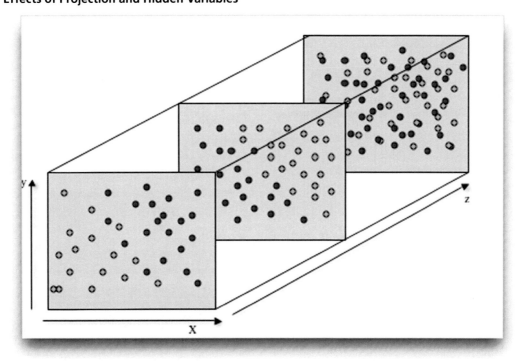

Linear Sensitivity Analysis

We first used simple linear regression on outcomes (the final ratio of oppositionals to extremists) to estimate the importance of parameters across the ensemble of scenarios. The length of the bars in a CARs graphic indicates how much the outcome changes in fractional terms if one varies a given parameter while averaging over the entire scenario ensemble and also over the different values of the parameter sample.[5] This and similar measures are used in many analysis programs, including Analytica and At Risk®. The bars measure correlation; they do not imply causality.

Figure 3.7 identifies four to six variables that appear to be important. We chose two of them, contagion rate and recovery rate, as the basis of axes for Figure 3.8. Some pattern is now discernible, but we would like to try looking at more than two inputs at once.

Figure 3.7
Linear Sensitivity of Final Ratio to Selected Parameters

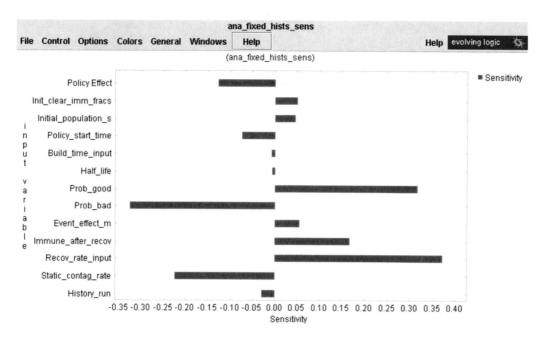

RAND *TR392-3.7*

---

[5] More precisely, in this CARs graphic, the length of the bars is proportional to the relevant coefficient in a linear-regression model fitting the scenario outcomes.

Figure 3.8
Choosing Better Axes Begins to Bring Out a Pattern

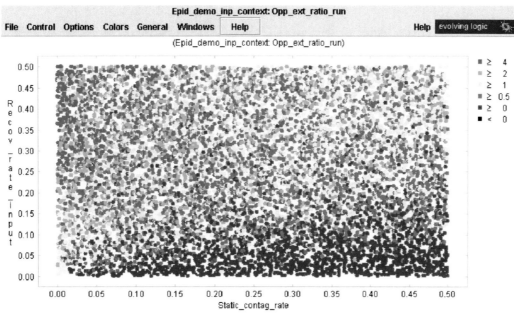

RAND *TR392-3.8*

Using Aggregation Fragments

The second method for examining, manipulating, and abstracting from exploratory analysis involves man-machine interaction. Here analyst insights are used to create hypotheses that can be adjudicated by inspecting visualizations. Since the clarity of two-dimensional plots varies with the choice of primary axes, the analyst can try to guess which particular combination of axes might reveal a strong pattern and can quickly test these guesses.

The guesses can be informed by knowledge of the problem. In particular, the analyst may postulate that *combinations* of low-level input parameters (e.g., dimensionless parameters) are important. In the current problem, one guess about how to display results would be to display them in a dot plot that uses the *ratio* of recovery rate to contagion rate as one axis.

Figure 3.9 uses that ratio for the y axis. It also uses an insight from the earlier linear sensitivity plot by plotting policy effectiveness on the x axis (where smaller numbers represent more-effective policies).

Whatever pattern might exist in Figure 3.9 is heavily obscured by dot-to-dot variation. The reason for this is what statisticians would call *interaction effects*. Ultimately, the problem is that we are trying to make sense of results in two-dimensional charts when the problem has more dimensions. Those "other" parameters create the noise.

Figure 3.10 performs some further abstraction by aggregating the results in Figure 3.9 into cells. This amounts to applying a kind of noise filter. If the plot region is divided into small cells, then the color assigned to a cell is the average color of the points within the cell. Results

Figure 3.9
Recovery-to-Contagion Ratio Versus Policy Effectiveness

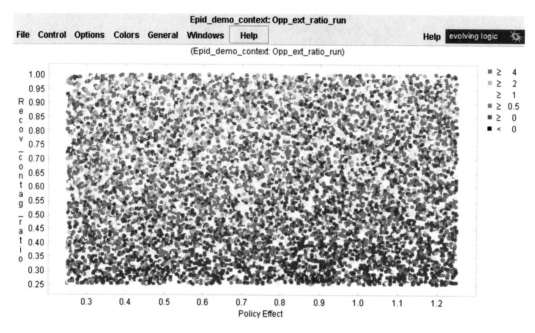

Figure 3.10
A "Region Plot"

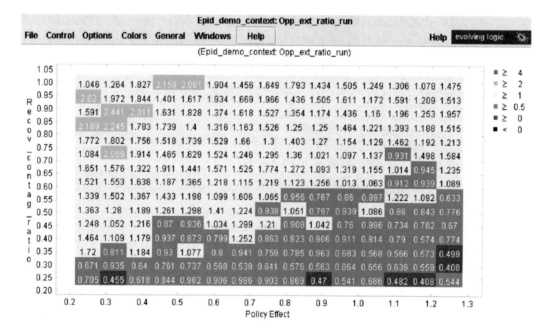

are indeed much sharper. Although there continue to be some peculiar features—as where outcomes are yellow, then green, then yellow again, moving horizontally—the general patterns are much more visible. Clearly, outcomes are bad if recovery rate divided by contagion rate is small (low on the chart), and they tend to be good if the recovery-to-contagion rate is large and policies are highly effective (top left).

Figure 3.11 shows a similar result, this time plotting recovery-to-contagion ratio versus immunity rates, rather than policy effect. As in Figure 3.10, the pattern is fairly clear but not entirely consistent.

A principal source of the remaining noise is the random exogenous events. Figure 3.12 shows what happens when we use the same output we used in Figures 3.9 and 3.10 but look at the mean results over the various future histories of such events. Now the figure has sharpened up a great deal, with few remaining inconsistencies. It follows that the ratio of recovery-to-contagion rates and the effect of policies go a long way in predicting outcomes. A relatively small number of such two-dimensional region plots can, in this problem, summarize the land-scape of outcomes obtained with MSG.

To summarize a full analysis of results from MSG, we might use variant charts in which the "other" parameters have fixed values and then juxtapose a number of such charts. For example, if only four parameters—v, x, y, and z—turned out to be important, a good summary analysis could be shown with four charts on a single page. These would chart v versus x if that were the richest single chart, but in successive charts in a row, y would have its low

Figure 3.11
A Region Plot of Final Ratio Versus Immunity Rates

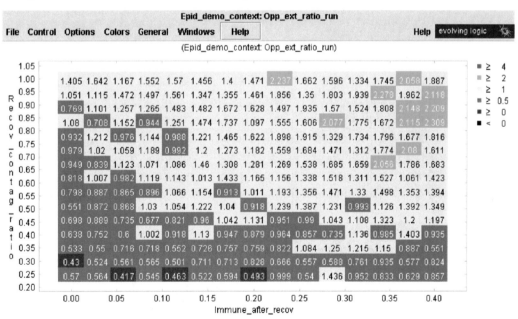

Figure 3.12
Averaging over the Stochastic Variations Sharpens the Pattern

Epid_demo_context: Opp_ext_ratio_avg

| File | Control | Options | Colors | General | Windows | Help | | Help | evolving logic |

(Epid_demo_context: Opp_ext_ratio_avg)

Recovery_contagion_ratio															
1.00	1.918	1.974	1.894	1.812	1.881	1.854	1.508	1.507	1.537	1.441	1.452	1.458	1.435	1.425	1.434
0.95	1.608	1.951	1.883	1.688	1.695	1.715	1.653	1.617	1.531	1.523	1.413	1.274	1.368	1.212	1.243
0.90	1.729	1.903	1.785	1.615	1.738	1.55	1.631	1.506	1.477	1.316	1.321	1.338	1.291	1.25	1.251
0.85	2.156	1.929	1.701	1.794	1.562	1.591	1.446	1.38	1.293	1.412	1.301	1.274	1.204	1.284	1.259
0.80	1.779	1.842	1.635	1.567	1.689	1.533	1.495	1.382	1.405	1.263	1.303	1.137	1.202	1.117	1.169
0.75	1.681	1.888	1.655	1.434	1.511	1.486	1.352	1.454	1.345	1.219	1.19	1.091	1.212	1.151	1.285
0.70	1.747	1.537	1.584	1.668	1.4	1.428	1.497	1.25	1.233	1.165	1.152	1.116	1.196	1.056	1.096
0.65	1.621	1.577	1.553	1.441	1.263	1.261	1.155	1.15	1.255	1.115	1.014	1.055	1.061	0.94	0.944
0.60	1.735	1.609	1.462	1.35	1.182	1.241	1.262	1.123	1.09	0.966	1.062	0.979	0.989	1.021	0.993
0.55	1.533	1.267	1.366	1.286	1.358	1.206	1.103	1.025	1.008	1.024	0.96	0.887	0.99	0.87	0.873
0.50	1.308	1.281	1.148	0.994	1.081	1.044	1.025	1.065	0.882	0.878	0.826	0.828	0.76	0.819	0.759
0.45	1.389	1.222	1.234	0.95	1.049	0.935	1.033	0.866	0.803	0.83	0.785	0.781	0.801	0.76	0.746
0.40	1.321	1.169	1.006	0.918	0.881	0.897	0.867	0.837	0.77	0.766	0.689	0.648	0.647	0.587	0.707
0.35	0.946	0.857	0.9	0.76	0.786	0.727	0.595	0.637	0.601	0.628	0.68	0.598	0.671	0.618	0.604
0.30	0.908	0.704	0.771	0.76	0.636	0.7	0.688	0.651	0.529	0.508	0.643	0.55	0.573	0.496	0.551

Policy Effect (x-axis: 0.2, 0.3, 0.4, 0.5, 0.6, 0.7, 0.8, 0.9, 1.0, 1.1, 1.2, 1.3)

Legend:
■ ≥ 4
■ ≥ 2
  ≥ 1
■ ≥ 0.5
■ ≥ 0
■ < 0

RAND *TR392-3.12*

and high values, and in successive charts in a column, z would have its low and high values, respectively. If more variables matter, as in the present case, parametric summary depictions would require more pages.[6]

Filters

Let us now turn to the third method, which involves filters. This method appeals to those without a good theory but with the task of making sense of complex results. As discussed in Chapter 2, it draws on research from the data-mining and cluster-analysis communities. We illustrate the approach using the Patient Rule Induction Method (PRIM) mentioned earlier (Friedman and Fisher, 1999), which searches through the ensemble of scenario outcomes to find, for example, those that are good. The result is an n-dimensional box in which most of the good cases are to be found and in which most outcomes are good (tuning parameters can change the criteria).

Research continues on such algorithms and, in our own work, on how to apply them in exploratory analysis, a rather different application from that for which they are typically developed.

Figure 3.13 shows the result of applying the PRIM algorithm to look for the good cases. One interesting aspect of the result is that good cases appear throughout the space defined by the ratio of recovery to contagion (y axis) and policy effects (x axis). Thus, terrorism can either

---

[6] This technique has been used in a number of defense applications (Davis, McEver, and Wilson, 2002).

Figure 3.13
Recovery-to-Contagion Ratio Versus Policy Effectiveness for Points Found in a PRIM Search for Good Outcomes

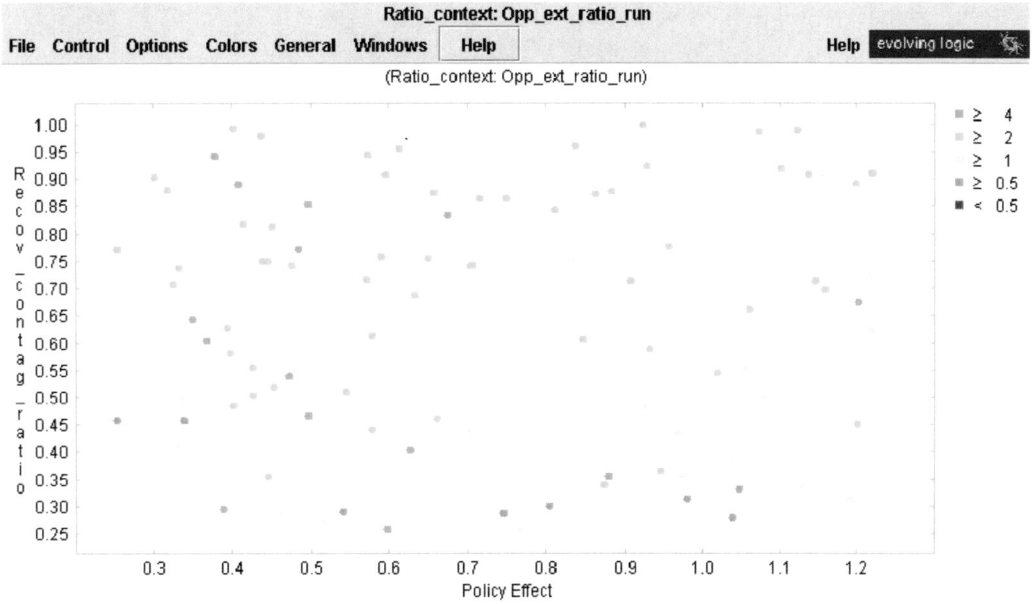

RAND *TR392-3.13*

wither away or be defeated even if policies are counterproductive (right side of the graphic, where the policy effect exceeds 1) or recovery rates are small (lower portion of the graphic). The good cases are also limited to certain ranges of the "other" parameters not plotted in the figure. If PRIM finds that good cases occur only within a narrow range of some parameter, then that parameter is important and is a good candidate for being used as the x or y axis.

Figure 3.14 shows that if one uses a better set of axes as suggested by PRIM (or by linear sensitivity analysis), order is more apparent, although some clutter still exists. In this case, the x axis is the fraction of those who recover from extremism who are subsequently immune.

Figure 3.15 illustrates again how noise filtering can substantially sharpen the conclusions. The results here are the same as those shown in Figure 3.14, except that averages are taken across scenarios within the cells shown. This certainly clarifies basic trends, but it has the disadvantage of hiding the substantial variation that occurs from one scenario to another, primarily because of the unpredictable exogenous effects.

To conclude this section, let us recall the significance of run-to-run variation—a point made earlier. Figure 3.16 compares the final oppositional-to-extremist ratio versus the average value of that ratio across the various stochastic samples of a given scenario. If, for example, the average outcome value is about 3, the range of values is from 0.5 to 8.5. That is, depending on the unpredictable exogenous effects, future history could vary from an explosion of extremists to their virtual elimination. Thus, when thinking about policy measures, one should think more of improving the odds than of moving with certainty. That would not be true if it turns

Figure 3.14
Results with Axes Suggested by PRIM

Ratio_context: Opp_ext_ratio_avg
File Control Options Colors General Windows Help
Help evolving logic
(Ratio_context: Opp_ext_ratio_avg) Good_Ratio22
Recov_contag_ratio
1.00 0.95 0.90 0.85 0.80 0.75 0.70 0.65 0.60 0.55 0.50 0.45 0.40 0.35 0.30 0.25
0.00 0.05 0.10 0.15 0.20 0.25 0.30 0.35 0.40
Immune_after_recov
≥ 4
≥ 2
≥ 1
≥ 0.5
< 0.5

RAND *TR392-3.14*

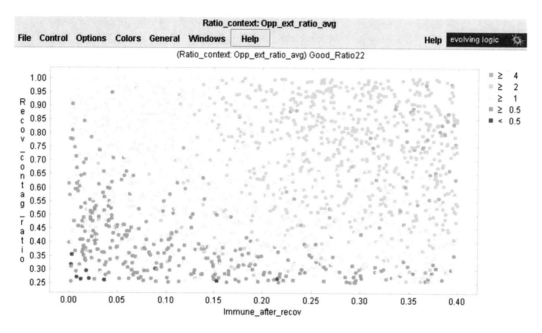

Figure 3.15
Noise-Filtered Results

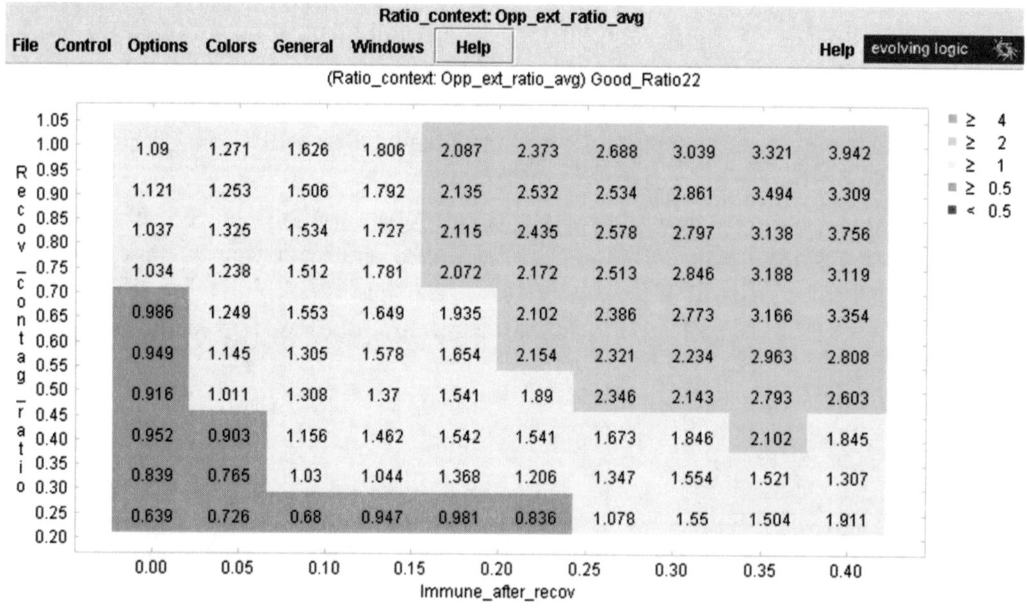

Ratio_context: Opp_ext_ratio_avg
File Control Options Colors General Windows Help
Help evolving logic
(Ratio_context: Opp_ext_ratio_avg) Good_Ratio22
Recov_contag_ratio
1.05 1.00 0.95 0.90 0.85 0.80 0.75 0.70 0.65 0.60 0.55 0.50 0.45 0.40 0.35 0.30 0.25 0.20
1.09 1.271 1.626 1.806 2.087 2.373 2.688 3.039 3.321 3.942
1.121 1.253 1.506 1.792 2.135 2.532 2.534 2.861 3.494 3.309
1.037 1.325 1.534 1.727 2.115 2.435 2.578 2.797 3.138 3.756
1.034 1.238 1.512 1.781 2.072 2.172 2.513 2.846 3.188 3.119
0.986 1.249 1.553 1.649 1.935 2.102 2.386 2.773 3.166 3.354
0.949 1.145 1.305 1.578 1.654 2.154 2.321 2.234 2.963 2.808
0.916 1.011 1.308 1.37 1.541 1.89 2.346 2.143 2.793 2.603
0.952 0.903 1.156 1.462 1.542 1.541 1.673 1.846 2.102 1.845
0.839 0.765 1.03 1.044 1.368 1.206 1.347 1.554 1.521 1.307
0.639 0.726 0.68 0.947 0.981 0.836 1.078 1.55 1.504 1.911
0.00 0.05 0.10 0.15 0.20 0.25 0.30 0.35 0.40
Immune_after_recov
≥ 4
≥ 2
≥ 1
≥ 0.5
< 0.5

RAND *TR392-3.15*

Figure 3.16
A Reminder That Scenario-to-Scenario Variation Is Very Large

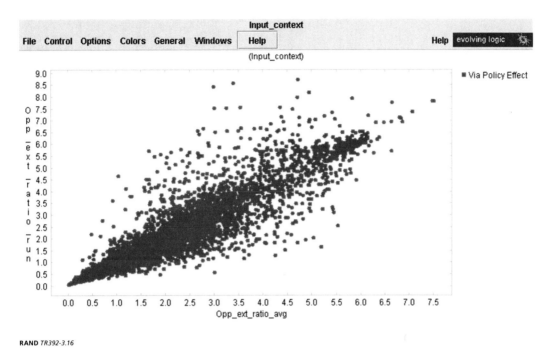

RAND *TR392-3.16*

out that the contagion rate is low, the recovery rate is high, policies work, and so on (right side of the chart). In such cases, outcome is predictably good and variance is low. Unfortunately, most scenarios are not of this type. *It follows that exploratory analysis is often about understanding variance rather than mean outcomes.*

Metamodeling

Let us now use the fourth method and seek a motivated metamodel that explains the scenario outcomes in understandable, albeit approximate, terms. In this case, we deliberately took a quick-and-dirty approach, rather than trying to be too clever. We reasoned quickly that the final ratio of oppositionals to extremists would probably grow in proportion to the recovery rate, inversely with the contagion rate, and so on, as shown in the equation below. We then left room for falsification by including an overall multiplier M and a linear sum of possible correction factors with unknown coefficients (the C's). The resulting postulated metamodel was

$$FinalRatio = C_0 + \frac{M(RR/CR)(1+FBI)}{CI(Infil)(EM)(Pbad/Pgood)}\{1 + C_1 RR + C_2 R + ...\}$$

where RR and CR are the recovery and contagion rates, FBI is the fraction of recovering people who become immune, CI is the effectiveness of the counterideology policy, Infil is the effectiveness of the infiltration policy, EM is the magnitude of the exogenous effects, and Pbad and

Pgood are the yearly probabilities of bad and good events, respectively. In the examples shown here, for simplicity, we hold the ratio Pbad/Pgood constant at a value of 2.

To our surprise, the crude motivated metamodel worked surprisingly well (Figure 3.16), although it required some correction factors (nonzero coefficients for some of the C's), indicating that the expression in curly brackets is not fully sufficient. The best-fit regression to the motivated metamodel was

$$FinalRatio = 0.92 - \left\{ 0.68 \frac{(RR/CR)(1+FBI)}{CI(Infil)(EM)(Pbad/Pgood)} \right\} Corr$$

$$Corr = 1 + 0.7\,RR - 2.5\,CR - 4.3\,FBI - 1.6\,CI - 1.7\,Infil - 0.32\,EM + 2.6\,Pbad + 1.3\,Pgood$$

The R^2 value for this metamodel is high, 0.76. The standard deviation of the difference between the actual final ratio and the one calculated from the metamodel is 0.26. It follows that one could argue that the equation defining the metamodel is a good approximate summary of what determines the final ratio of oppositionals to extremists. Thus, it both predicts and explains— especially to the extent that the correction factor is relatively small.

In this case, the correction factor is neither large nor small enough to ignore. One way to see that is to find the best simplified motivated metamodel, one that has only a constant term and a scale factor (i.e., it assumes C1, C2, . . . are zero). The result in that case is

$$FinalRatio = 1.3 + 0.50 \left\{ \frac{(RR/CR)(1+FBI)}{CI(Infil)EM(Pbad/Pgood)} \right\}$$

with an R^2 of 0.35 and a standard deviation of ratio minus predicted of 0.4. Thus, the simple model goes a fair way in explaining results, but—not surprisingly—it does not go far enough.[7] Perhaps this would be a good preface to any discussion of the corrected model.

It is important to note that we could have developed an even better motivated metamodel by an iterative cycle of observation, thinking, model revision, and observation. We did not do so, however, because we had wanted to do the experiment without inserting deep knowledge or requiring significant mathematical work.

The results from the full model are compared in Figure 3.17 with those from the motivated metamodel.

---

[7] To understand why the corrections are necessary would require considerably more thought, and with such thought, it would very likely prove desirable to change the basic form (i.e., the form within the curly brackets). Our purpose, however, was merely to illustrate how relatively straightforward application of the motivated metamodeling approach could be useful.

Figure 3.17
Comparison of Results from Full Model and Motivated Metamodel

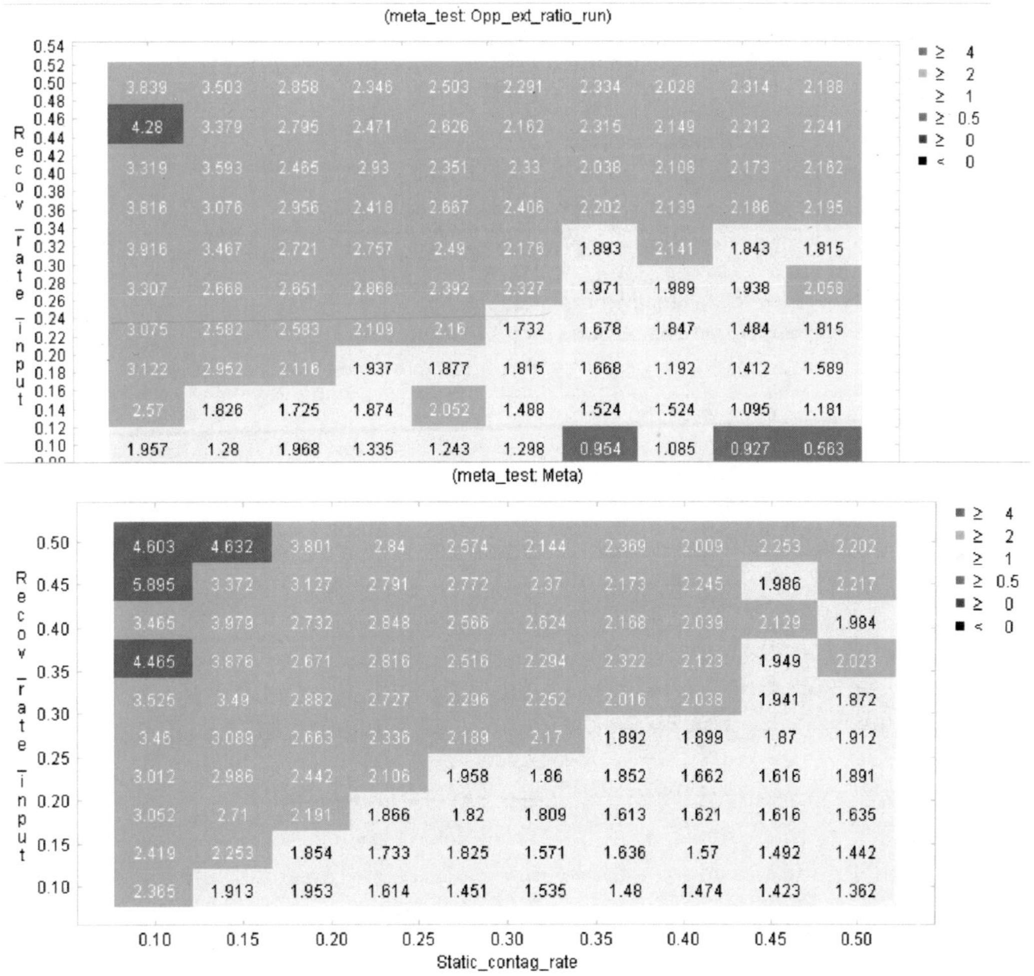

(meta_test: Opp_ext_ratio_run)

	0.10	0.15	0.20	0.25	0.30	0.35	0.40	0.45	0.50
0.50	3.839	3.503	2.858	2.346	2.503	2.291	2.334	2.028	2.314	2.188
0.45	4.28	3.379	2.795	2.471	2.626	2.162	2.315	2.149	2.212	2.241
0.40	3.319	3.593	2.465	2.93	2.351	2.33	2.038	2.108	2.173	2.182
0.36	3.816	3.076	2.956	2.418	2.687	2.406	2.202	2.139	2.186	2.195
0.32	3.916	3.467	2.721	2.757	2.49	2.176	1.893	2.141	1.843	1.815
0.28	3.307	2.668	2.651	2.868	2.392	2.327	1.971	1.989	1.938	2.058
0.24	3.075	2.582	2.583	2.109	2.16	1.732	1.678	1.847	1.484	1.815
0.18	3.122	2.952	2.116	1.937	1.877	1.815	1.668	1.192	1.412	1.589
0.14	2.57	1.826	1.725	1.874	2.052	1.488	1.524	1.524	1.095	1.181
0.10	1.957	1.28	1.968	1.335	1.243	1.298	0.954	1.085	0.927	0.563

Recov_rate_Tinput (vertical axis): 0.54, 0.52, 0.50, 0.48, 0.46, 0.44, 0.42, 0.40, 0.38, 0.36, 0.34, 0.32, 0.30, 0.28, 0.26, 0.24, 0.22, 0.20, 0.18, 0.16, 0.14, 0.12, 0.10, 0.09

Legend: ≥ 4, ≥ 2, ≥ 1, ≥ 0.5, ≥ 0, < 0

(meta_test: Meta)

	0.10	0.15	0.20	0.25	0.30	0.35	0.40	0.45	0.50
0.50	4.603	4.632	3.801	2.84	2.574	2.144	2.369	2.009	2.253	2.202
0.45	5.895	3.372	3.127	2.791	2.772	2.37	2.173	2.245	1.986	2.217
0.40	3.465	3.979	2.732	2.848	2.566	2.624	2.168	2.039	2.129	1.984
0.35	4.465	3.876	2.671	2.816	2.516	2.294	2.322	2.123	1.949	2.023
0.30	3.525	3.49	2.882	2.727	2.296	2.252	2.016	2.038	1.941	1.872
0.25	3.46	3.089	2.663	2.336	2.189	2.17	1.892	1.899	1.87	1.912
0.20	3.012	2.986	2.442	2.106	1.958	1.86	1.852	1.662	1.616	1.891
0.15	3.052	2.71	2.191	1.866	1.82	1.809	1.613	1.621	1.616	1.635
	2.419	2.263	1.854	1.733	1.825	1.571	1.636	1.57	1.492	1.442
0.10	2.365	1.913	1.953	1.614	1.451	1.535	1.48	1.474	1.423	1.362

Recov_rate_Tinput (vertical axis): 0.50, 0.45, 0.40, 0.35, 0.30, 0.25, 0.20, 0.15, 0.10

Static_contag_rate (horizontal axis): 0.10, 0.15, 0.20, 0.25, 0.30, 0.35, 0.40, 0.45, 0.50

Legend: ≥ 4, ≥ 2, ≥ 1, ≥ 0.5, ≥ 0, < 0

Conclusions

It was not our purpose to draw firm conclusions from this experiment, which was not a full analysis, but some tentative conclusions emerged (see Table 3.2). Some of the conclusions were arguably not obvious, except in retrospect. For example, policies that can prove to be counterproductive despite expectations are especially risky if—in the absence of those policies—the "natural" lifetime of the extremism disease is short.[8] Also, for thinking about what subpopula-

---

[8] In a sense, this is just a special case of the admonition that policymakers should try to think ahead, rather than being driven to actions based on what is apparent "now." The admonition, however, is very difficult to follow. Simulations such as the epidemiological model can help a good deal. The system dynamics literature has made this point for some years (Sterman, 2000).

Table 3.2
Observations from Prototype Analysis

Non-obvious scenarios arising from MSG

Scenario-to-scenario variation of trajectories is huge—the future is not reliably predictable:

    In many cases, extremism dies off naturally.

    In other cases, extremism "takes off" and becomes dominant.

    Outcomes tend to cluster at good and bad extremes.

    Close sequences of exogenous events have big effects.

Insights

Ratio of recovery rate to contagion rate is significant.

Ratio of probabilities for bad and good events is significant.

Immunity-rate assumptions are significant: Those who become immune reduce the pool of those who can become infected subsequently. Targeting potential or recent recoverers is therefore valuable.

Policies that can be counterproductive should be a special concern if there is reason to believe the "natural" lifetime of contagion will be modest.

Policies that might otherwise have significant effects may have no effect if they occur at the wrong time—with "bad" exogeneous events wiping out their influence.

---

tions to target with counterideology campaigns, the analysis suggests that we should not focus unduly on the uncommitted (oppositionals and passives), because the fraction of those who become immune after recovery also matters significantly.

In this experiment based on a reasonably good model, but one that had not yet been formally validated or peer-reviewed, there was a basis for assessing the credibility of the assumptions. Because there is a meaningful story supported by an understandable causal model, conclusions can be assessed and sometimes judged credible "on their face." Where a conclusion is less obviously credible, one can go back to the model and work through the relevant scenarios in more detail as necessary. Thus, exploratory analysis amidst great uncertainty *can* generate legitimate insights, even though no one can rigorously validate the model or any set of specific assumptions.[9]

By and large, the experiment demonstrated well that MSG, when used with a reasonably good model that has incorporated profound uncertainties, can indeed yield many results and insights that would not be obtained from more normal exercising of the model. Furthermore, because we approached the challenge within the MSG paradigm of wanting to explore as much of the possibility space as we could conceive, we found ourselves stretching our minds and substantially modifying the model itself. In particular, the recognition that large "exogenous events" were likely to be both important and unpredictable materially affected the model and our analysis.

---

[9] How to broaden the concept of validation for such highly uncertain models is currently being discussed in a National Academy of Sciences panel on modeling and simulation to which one of the authors (Davis) has contributed (Davis and Bigelow, 1998).

4. Experiment Two: Exploratory Analysis Starting Without a Model

The Starting Point: Constructing an Initial Model

As noted earlier, we began our next-nuclear-use (NNU) experiment by writing a report characterizing the circumstances of the NNU in a number of dimensions. The first question was whether using that report as the model would lead to useful MSG. After experimentation, we concluded that it would not, primarily because the dimensions provided were static attributes that said little about significance or implications for the future. The report itself did have considerable insightful discussion, but a useful model for scenario generation would need to capture and extend the reasoning used by the author, not merely work from the list of dimensions.

We also observed that the cases that could be generated from the list of dimensions were not really scenarios in any usual sense—they merely described situational attributes, without directly conveying a sense of context, dynamics, or "story." Interpretation and sense-making depend on tacit knowledge. That is, if we thought one of the combinations generated was interesting or uninteresting, it was because of nonexplicit assumptions. We considered that quite unsatisfactory. Thus, we turned to an approach of moving from our starting point to something more nearly like a causal model. The approach we described at the beginning of this report was an outgrowth of this experience.

Our next step was to embellish the original list of dimensions with items addressing missing points that we considered essential. These included context (why the NNU had occurred), what would happen next (additional attacks, counterattacks, etc.), and measures of outcome that we called "seriousness" (to both the United States as a whole and the intelligence community more narrowly). We do not present results of that exercise here because we eventually concluded that it would be a more useful experiment if we put aside the fact that we would have approached the entire problem differently, and instead went back to the full report, treating it as analogous to what might have emerged from a strategic-planning exercise or a brainstorming session. We then studied the report's text for substantive knowledge not captured by the initial list of dimensions. By inferring and extrapolating from the textual material of the report, we were able to find a considerable, albeit implicit, conceptual structure. We turned that implicit structure into the beginnings of a dynamic model in Analytica, as shown in Figure 4.1. The figure includes the highlighted dimensions of the report (in blue) and also U.S. policies, both before and after the NNU (in green), intermediate variables more directly relevant to some of

Figure 4.1
A Conceptual Model of Next Nuclear Use

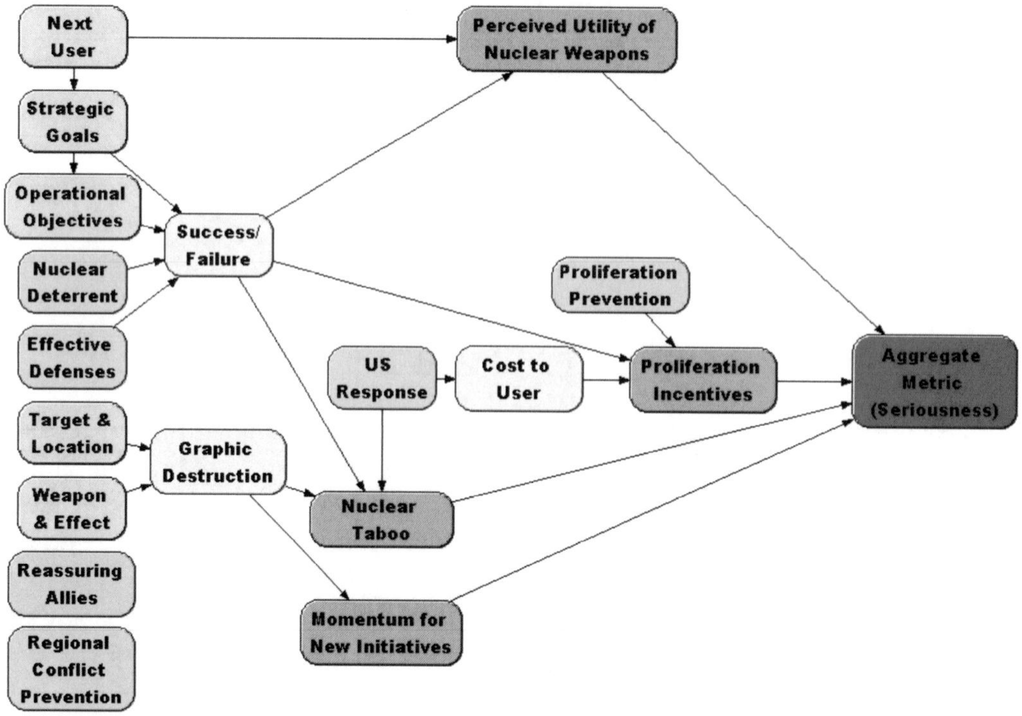

RAND *TR392-4.1*

the intelligence community's concerns (in yellow), and at least four major, longer-term effects arising from the NNU (in orange). These effects could serve as metrics, allowing us to classify and compare many scenarios simultaneously. We could also combine these four concerns into a single, aggregate scenario metric (in red).

New Methods for Dealing with Profound Uncertainty in the Models

Our first-cut model added conceptual significance to the scenarios, but at a potential cost: If we used the model for MSG, we could obviously not obtain scenarios that violated model assumptions. Consider just one fragment of an influence diagram, one that asserts that more of A leads to more of B. In real-world problems involving complex social systems, any such influence-diagram fragment is typically merely an assertion about what is allegedly normal or perhaps intended. Even if experts agreed upon the fragment, they would be agreeing only on what they thought was "usually" a good approximation. In other cases, the fragment might represent only a majority view. That is not sufficient for our purposes. The models we use for MSG must recognize uncertainties, including structural uncertainties, and must accommodate the possibility of competing models.

Suppose, for example, that A represented a proposed policy intended to improve the situation (increase B). In the real world, policies often prove counterproductive. That is, in influence-diagram terms, we do not really know the directionality or sign of the arrow. Also, whereas it is usual to assume that a single algorithm relates B to A, so that each instance relating to A would have the same effect on B, in the real world, the effects might be different in magnitude and character, even if not in directionality. Thus, a game or brainstorming session might assert that a strong U.S. response to the NNU would always decrease subsequent proliferation. In reality, the same response would probably have a number of different effects, depending on contextual details (including essentially random factors).

We responded to this dilemma by translating influence-diagram relationships into uncertain model relationships. That is, when performing MSG, we allowed for the possibility of reversed directionality and of diverse effects. To our knowledge, this is quite unusual. One could say that we generated scenarios with many different models, such as might be proffered by experts with different opinions or experts saying, "Well, it might be this way or it might be that way."

To illustrate how we proceeded at a technical level, consider the relationship between the next-use weapon and its potential destructive effect, displayed in Figure 4.2 on a 0-to-9 scale. To let any weapon have any effect would simply make weapon choice irrelevant in the final reckoning. On the other hand, assigning specific effect sizes to weapon types would ignore cases where a smaller attack is "lucky" or a larger attack somehow falters.

Our structure is displayed in Figure 4.2 in a simplified form. After the weapon type is selected (via a random process), the effect size is allowed to vary randomly within a range set by the weapon type. As indicated on the left side of the figure, an improvised weapon might vary in effect from 0 (a dud or fizzle) to 5 (moderate-to-heavy damage). A large-weapon attack,

Figure 4.2
Using Stochastic Methods to Reflect Structural Uncertainty

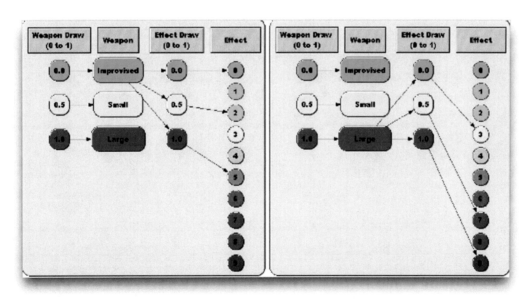

RAND *TR392-4.2*

on the other hand (right side of the figure), might vary in effect from 3 (limited damage) to 9 (severe damage).

Using this structure, as the value of a single variable (Effect Draw) increases, we can move from the best plausible case (= 0, in green) to the worst plausible case (= 1, in red), given the scenario elements selected thus far.

To construct Figure 4.3, we used random draws to flip the valence of influence arrows (from positive to negative, or vice versa). In the context of U.S. policy, this meant letting policy effects vary from highly effective (multiplier = 0.25, in green) to exactly ineffective (multiplier = 1, in orange) to strongly counterproductive (multiplier = 1.50, in red).

Figure 4.3 also shows how we might add significance to these ranges by linking them to policy resource levels. On the left, a richly resourced policy can vary from strongly effective (0.25) to moderately counterproductive (1.25). On the right, a poorly resourced policy, even at its best, can be only moderately effective (multiplier = 0.5); at its worst, it will be strongly counterproductive (1.50).[1]

We emphasize here that it is the range of effects, and not the particular percentage of cases that fall one way or another, that is of use for MSG. We are interested in covering the plausible scenario space with minimal weighting by likelihood; in fact, the unlikely (but still plausible) corners of the scenario space may yield the most critical lessons. Therefore, while the right-hand side of Figure 4.3 implies that 50 percent of scenarios will be counterproductive, this does not reflect an expert judgment of probabilities. Instead, it is simply the mathematical result of varying cases uniformly between a plausible best case and worst case. Similarly, in looking at results of MSG, we should be more interested in noting the circumstances of

Figure 4.3
Using Stochastic Methods to Reflect Structural Uncertainty, Allowing Policy Effects to Vary

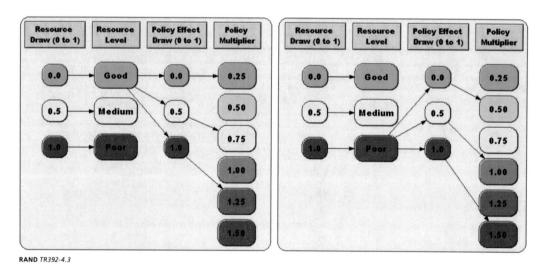

RAND TR392-4.3

---

[1] In this model, policies reduce "problems" through the formulation that problem values = (potential problem) x (policy multiplier). Therefore, a policy multiplier of 0.25 eliminates 75 percent of a potential problem, while a policy multiplier of 1.50 worsens the potential problem by 50 percent.

predominantly good results (or predominantly bad results) than in precisely estimating the fraction of results, for a given situation, that are good.

Textual Stories and Visualizations from the MSG Experiment

Before moving on to a visual display of outcomes from our MSG experiments on the NNU problem, we demonstrate in Table 4.1 that our methodology was able to generate textual "stories" automatically. Seven examples of such stories are shown.

On a macro level, these textual stories have limited value. Not only would it take an extremely long time to read 10,000 of them, it is even more difficult to extract broad lessons from such a list. We preferred instead to use the CARs system to generate visual displays and then to operate on the data in various ways in pursuit of meaningful patterns.

In the figures below, we again use some of the techniques discussed earlier. Recall that by feeding the rough-cut conceptual model into CARs, we can explore thousands of cases. We might want to explore the entire scenario space, varying all situational attributes; or we might fix certain attributes and let others vary. We can then evaluate these scenarios against any outcome of interest. To illustrate, we focused on the question, How might revenge-motivated attacks by the Democratic People's Republic of Korea (DPRK) affect the nuclear taboo?

Figure 4.4 presents a first-cut visualization from the NNU experiment. Each dot represents one of the 10,000 scenarios generated. The axes are scenario dimensions, and the colors indicate levels of the nuclear taboo. High/red values indicate that the nuclear taboo has been weakened, while low/green values indicate a continued, strong taboo.

Table 4.1
Textual "Stories"

Run	Story
1	To serve its vision by inflicting mass deaths, a nonstate group targets leadership in a neighboring country, early in the conflict, using an improvised weapon. The weapon causes heavy damage and achieves its goal.
248	To defend itself by deterring its adversary, Iran targets military nodes in a nonneighboring country, mid-war, using a small weapon. The weapon causes heavy damage and achieves its goal.
1,481	To end a war by preempting nuclear forces, Pakistan targets nuclear assets in a neighboring country, early in the conflict, using a small weapon. The weapon is a dud and fails to achieve its goal.
5,225	To limit damage by preempting nuclear forces, Israel targets nuclear assets in a nonneighboring country, late in the war, using 11 to 20 weapons larger than 5 KT. The weapons cause heavy damage and fail to achieve the goal.
7,311	To seek revenge by inflicting mass deaths, the Democratic People's Republic of Korea (DPRK) targets civilians in the United States during a crisis, using a small weapon. The weapon causes heavy damage and achieves its goal.
9,464	Without authorization, Pakistan targets no specific target in a nonneighboring country, early in the conflict, using a small weapon. The weapon fizzles.
10,000	To limit damage by paralyzing decisionmaking, China targets military nodes in a neighboring country, early in the conflict, using a large weapon. The weapon causes severe damage and fails to achieve its goal.

Figure 4.4
Exploring Revenge Attacks by DPRK

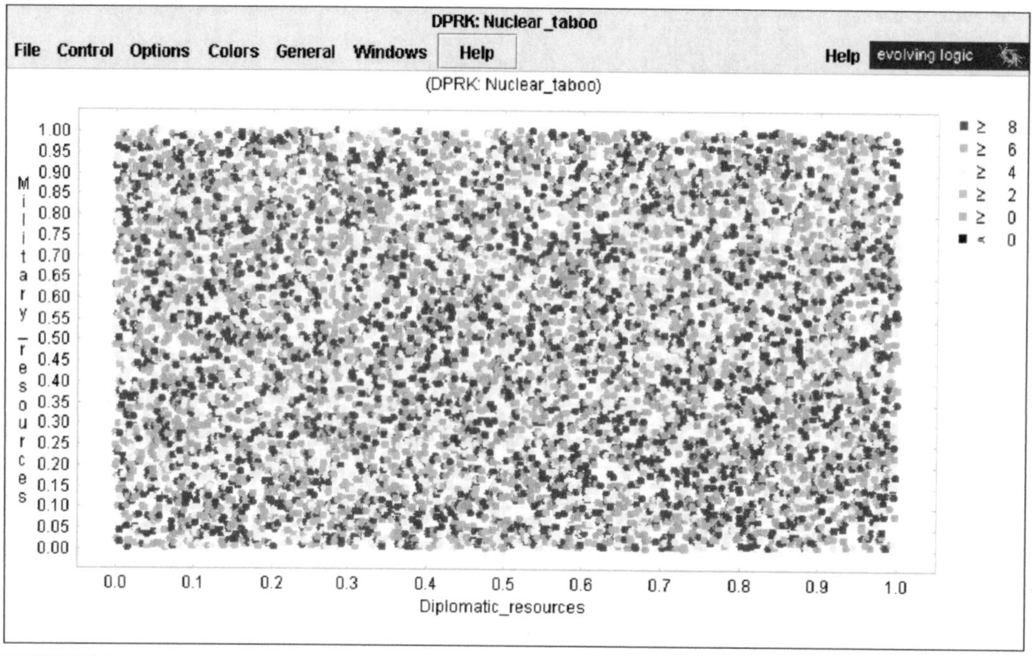

RAND *TR392-4.4*

In this first cut, naively choosing two axes (U.S. military and diplomatic resources), we see no pattern across the 10,000 cases.

Figure 4.5 shows results of applying the linear sensitivity method. Although such results must be interpreted with caution, they can suggest which variables warrant further exploration or visualization. Five variables appear to be important in relation to the strengthening or weakening of the nuclear taboo:

1. The destructiveness of the attack
2. Any follow-on attacks on the target country
3. The effectiveness of U.S. punishment
4. The specific targets chosen by DPRK
5. DPRK's operational objective in seeking revenge

We could explore each pair of these variables in ten visualizations, but we show merely one example here.

One of the ten visualizations is displayed in Figure 4.6. Again, color indicates the long-term effect of the NNU on the nuclear taboo, with red indicating a bad effect. The dots here are the same 10,000 scenarios shown earlier, but with different axes a pattern has begun to emerge. The visualization suggests that the taboo is weakened most when the United States does not effectively punish DPRK and when the level of destructiveness is in a mid-range.

Figure 4.5
Linear Sensitivity Analysis for Revenge-Attack Cases

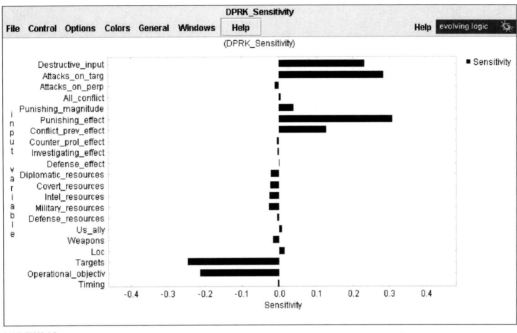

RAND *TR392-4.5*

That is, effects vary nonmonotonically with attack destructiveness. This nonmonotonicity is the type of unusual structure in graphical results of exploratory analysis that often signals an interesting phenomenon or potential insight to be gained.

The reason for this result is that if destructiveness is low, DPRK has failed in its goal of revenge. This failure keeps the taboo in check. When destructiveness is high, the outcry as the result of widespread damage and fatalities may also keep the taboo in check. Only when destructiveness is high enough for DPRK to claim a success but not so high that the rest of the world is outraged will the taboo be severely weakened.

Was this a *new* insight from the prototype MSG? This raises a deep issue. As with human gaming and simulation generally, one often learns something insightful that afterward—but not before—seems intuitively obvious. Further, others participating in the same activity may have already had the particular insight. Or some experts who had been thinking about the issues in depth over a period of years might take it for granted. Nonetheless, for those who see the point for the first time, the exercise has been insightful.

Beyond that, recall that the model used in our prototype experiment was fairly primitive (and not what we would have built if we were approaching the NNU problem ourselves). Nonetheless, MSG revealed some interesting patterns. More would likely emerge from in-depth work of the same kind.

Figure 4.6
A Dot Plot for Revenge Attacks by DPRK

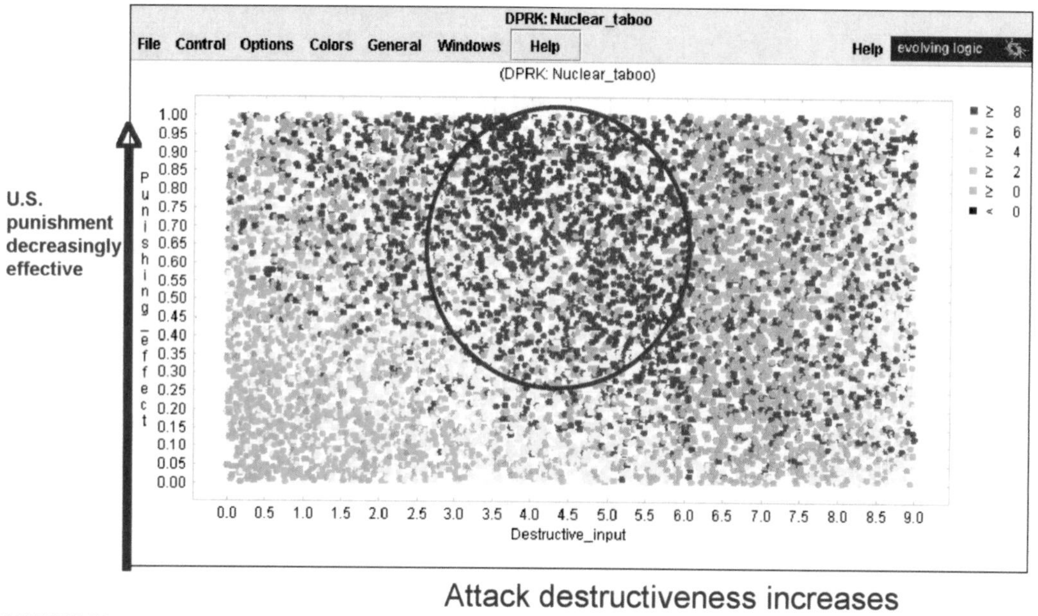

Attack destructiveness increases

RAND *TR392-4.6*

Lessons Learned from the NNU Experiment

We shall not elaborate here on the various negative lessons learned from our experiment which took us down paths that proved fruitless; rather, we emphasize the more-positive lessons.

First, the experiment strongly confirmed the notion that one should take a model-building approach from the outset. Doing so materially affects all aspects of the activity, including initial games, brainstorming, and strategic-planning exercises. The approach, however, should take a broad view of "model" and take extreme measures to recognize and assimilate uncertainties. Causality is essential to the generation of stories, explanations, and insights, but causality relationships are often difficult to uncover and, even in principle, are ambiguous when one is dealing with complex social systems.

With this attitude in mind, a subject area could be studied over time with a sequence such as the following:

1. At least some prior thinking and structuring of the problem
2. Initial brainstorming or gaming
3. Building a minimal causal model with structural uncertainties (akin to building multiple causal models)
4. Further gaming or human interactions
5. Further movement toward a well-conceived, rich, and highly uncertainty-sensitive system model

We see MSG and its interpretation as valuable at multiple points along the way, not only for analysts, but also for human groups interacting with "data" and offering new hypotheses and new pieces of information that had previously been only tacit.

The ideal endpoint for the process would be to achieve a system model that is comprehensive, structured, rich, and able to explain scenario dynamics, while reflecting the full range of appropriate uncertainties. That is, the ideal endpoint would be not an answer machine (a deterministic model and best-estimate data), but rather a sound model for at-the-time MSG applications that could help guide planning under uncertainty.

We conclude this chapter with a caution: Although the tools and methodology we have described could prove quite powerful in this type of work, where one begins without the luxury of a good model, success will depend not only on the experts used in games and the like, but also on the subject-area knowledge and talent of the analysts.

5. Conclusions

The research described here was a learning experience despite our considerable past experience with exploratory analysis, exploratory modeling, and related tools. The test problems provided new challenges, such as the need to allow for substantial "structural uncertainty" in models (unknown unknowns) and the need to allow for frequent instances in which we do not know even the directionality of effects.

We concluded that to do justice to some of these issues, agent-based modeling would be necessary in future work, and in some applications, other types of model (e.g., Bayesian nets or influence nets) might be quite useful.

We also concluded that the tools for MSG are pushing at the frontiers of knowledge and technology. Although we were able to accomplish a great deal, we found it necessary to use workarounds and to rethink what was needed. We would expect knowledge and know-how in this domain to continue building for some years, based on the timescales on which analogous new methods have developed.

We also have recommendations for those interested in pursuing MSG. The first is that one should have a clear objective when using MSG. We recommend an emphasis on suggesting ways to improve strategy (e.g., hedging) and to prepare for possible challenges.

We recommend further that MSG be approached with the explicit intent of developing a "model," rather than using unstructured or essay-style knowledge. We have in mind, to be sure, different kinds of model than are often thought of.

Next, we recommend two parallel tracks of further work. First, sponsors would be well served by commissioning serious "system model" work in one or a few areas of continuing interest. This would lead to richer and more coherent models, which could be exploited by MSG. Such models should include agent-based subcomponents or abstractions based on experiments with agent-based models.

The second track should build on the vision of working from initial, rough problem formulation, followed by initial modeling, brainstorming and gaming, and iterative accretion of causal model features until, at some point, enough is understood to reprogram into something more coherent *or,* by analogy with experience in human gaming, enough has been learned to declare success and move on, without bothering to tidy up and carefully check the model itself.

We believe that investments should be made in both theory and tools for MSG and its use. There are many deep issues involved that merit being thought through. Theoretical work could avoid or mitigate severe problems of misinterpretation in MSG and could guide tool development. That said, tool development can often lead the way—people often learn by doing. Thus, tool development and application should proceed in parallel with theory and should interact with it.

Our final recommendation is to move directly to real applications. The theory and tools are far enough along to justify that, and history tells us that working real problems is a powerful stimulus to the development and refinement of methods and tools.

We also have specific suggestions for tool-investment priorities:

Tools for Scenario Generation and Exploration

- Seamless integration of models into the CARs environment
- Mechanisms for rapid insertion of diverse models into CARs
- Better mechanisms for exploring across model *structures*
- Better mechanisms for building in and exploiting multiresolution modeling, including motivated metamodeling

Graphics and Visualization

- Additional and improved interfaces and display options

Analysis

- Review of data-mining algorithms for use in discovering classes of scenario
- Review or building of search algorithms to find scenarios with specific properties

Although models written in java, C++, and Excel connect readily to CARs, it is currently tedious to connect Analytica models to CARs. That connection could be made much more seamless, even with sophisticated Analytica models, which would be quite convenient because of Analytica's value in building and developing influence-diagram models. CARs could also be given more mechanisms both for exploring across different model *structures* and for building in multiresolution features, which can be critical for sense-making.

It is worth noting here that there are general issues associated with connecting good modeling platforms with platforms for scenario generation, analysis, and visualization (such as CARs). Seldom can one platform do everything well. Analytica, for example, has excellent features for influence diagrams, so-called "intelligent arrays" that are quite powerful for qualitative modeling as well as more traditional work, and substantial statistical methods. Other modeling platforms, such as iThink,™ GoldSim,™ and Hugin, would also be good candidates for experimentation.

Although CARs graphics and visualizations are already good, many improvements are possible, especially as experience builds about what different classes of analysts and decision-makers find most useful.

At the level of analysis, we see a substantial potential for improved methods of manipulation, filtering, and abstraction—particularly by drawing upon and adapting state-of-the-art methods in such related fields as data-mining. Some of these, such as the technique we illustrated for finding good or bad cases, are best regarded as "search" techniques.

References

Bankes, Steven C., "Exploratory Modeling for Policy Analysis," *Operations Research*, Vol. 41, No. 3, 1993, 435–449.

Banks, H. T., and C. Castillo-Chavez (eds.), *Bioterrorism—Mathematical Modeling Applications in Homeland Security*, Philadelphia, PA: Society for Industrial and Applied Mathematics (SIAM), 2003.

Botterman, Maarten, Jonathan Cave, James P. Kahan, and Neil Robinson, *Foresight Cyber Trust and Crime Prevention Project: Gaining Insight from Three Different Futures*, London: Office of Science and Technology, 2004.

Builder, Carl H., *Toward a Calculus of Scenarios*, Santa Monica, CA: RAND Corporation, N-1855-DNA, 1983. As of April 16, 2007:
http://www.rand.org/pubs/notes/N1855/

Castillo-Chavez, Carlos, and Baojun Song, "Models for the Transmission Dynamics of Fanatic Behaviors," chap. 7 in H. T. Banks and C. Castillo-Chavez (eds.), *Bioterrorism: Mathematical Modeling Applications in Homeland Security*, Philadelphia, PA: Society for Industrial and Applied Mathematics (SIAM), November 2003.

Cohen, William, *Quadrennial Defense Review Report*, Washington, DC: Department of Defense, 1997.

Darlington, Richard B., "Multivariate Statistical Methods." Last accessed in 2004:
http://comp9.psych.cornell.edu/Darlington/factor.htm

Davis, Paul K., "Institutionalizing Planning for Adaptiveness," in Paul K. Davis (ed.), *New Challenges for Defense Planning: Rethinking How Much Is Enough*, Santa Monica, CA: RAND Corporation, 1994a, 51–73. As of April 16, 2007:
http://www.rand.org/pubs/monograph_reports/MR400/

_____, "Protecting the Great Transition," in Paul K. Davis (ed.), *New Challenges for Defense Planning: Rethinking How Much Is Enough*, Santa Monica, CA: RAND Corporation, 1994b. As of April 16, 2007:
http://www.rand.org/pubs/monograph_reports/MR400/

_____, *Analytic Architecture for Capabilities-Based Planning, Mission-System Analysis, and Transformation*, Santa Monica, CA: RAND Corporation, 2002a. As of April 16, 2007:
http://www.rand.org/pubs/monograph_reports/MR1513/

_____, "Synthetic Cognitive Modeling of Adversaries for Effects-Based Planning," *Proceedings of the SPIE*, Vol. 4716, No. 27, 2002b, 236–250.

Davis, Paul K., and James H. Bigelow, *Experiments in Multiresolution Modeling (MRM)*, Santa Monica, CA: RAND Corporation, MR-1004-DARPA, 1998. As of April 16, 2007:
http://www.rand.org/pubs/monograph_reports/MR1004/

_____, *Motivated Metamodels: Synthesis of Cause-Effect Reasoning and Statistical Metamodeling*, Santa Monica, CA: RAND Corporation, MR-1570-AF, 2003. As of April 16, 2007:
http://www.rand.org/pubs/monograph_reports/MR1570/

Davis, Paul K., James H. Bigelow, and Jimmie McEver, *Exploratory Analysis and a Case History of Multiresolution, Multiperspective Modeling,* Santa Monica, CA: RAND Corporation, RP-925, 2001. As of April 16, 2007:
http://www.rand.org/pubs/reprints/RP925/

Davis, Paul K., Michael Egner, and Jonathan Kulick, *Implications of Modern Decision Science for Military Decision-Support Systems,* Santa Monica, CA: RAND Corporation, 2005. As of April 16, 2007:
http://www.rand.org/pubs/monographs/MG360/

Davis, Paul K., Jimmie McEver, and Barry Wilson, *Measuring Interdiction Capabilities in the Presence of Anti-Access Strategies: Exploratory Analysis to Inform Adaptive Strategies for the Persian Gulf,* Santa Monica, CA: RAND Corporation, MR-1471-AF, 2002. As of April 16, 2007:
http://www.rand.org/pubs/monograph_reports/MR1471/

de Mesquita, Bruce Bueno, *The War Trap,* New Haven, CT: Yale University Press, 1983.

Dewar, James, *Assumption Based Planning,* Cambridge, UK: Cambridge University Press, 2003.

Dewar, James, James Gillogly, and Mario Juncosa, *Non-monotonicity, Chaos and Combat Models,* Santa Monica, CA: RAND Corporation, R-3995-RC, 1991. As of April 16, 2007:
http://www.rand.org/pubs/reports/R3995/

Dobbins, James, John G. McGinn, Keith Crane, Seth G. Jones, Rollie Lal, Andrew Rathmell, Rachel M. Swanger, and Anga Timilsina, *America's Role in Nation-Building: From Germany to Iraq,* Santa Monica, CA: RAND Corporation, 2003. As of April 16, 2007:
http://www.rand.org/pubs/monograph_reports/MR1753/

Epstein, Joshua M., and Robert L. Axtell, *Growing Artificial Societies: Social Science from the Bottom Up,* Cambridge, MA: MIT Press, 1996.

Forrester, Jay Wright, *Urban Dynamics,* Cambridge, MA: Wright Allen Press, 1969.

Friedman, J., and N. Fisher, "Bump Hunting in High-Dimensional Data," *Statistics and Computing*, Vol. 9, No. 2, 1999, 1–20.

Ghamari-Tabrizi, Sharon, *The Worlds of Herman Kahn: The Intuitive Science of Thermonuclear War,* Cambridge, MA: Harvard University Press, 2005.

Holland, John H., and Heather Mimnaugh, *Hidden Order: How Adaptation Builds Complexity,* New York: Perseus Publishing, 1996.

Ilachinski, Andrew, *Artificial War: Multiagent-Based Simulation of Combat,* Singapore: World Scientific Publishing Company, 2004.

Kahn, Herman, *On Thermonuclear War,* Princeton, NJ: Princeton University Press, 1966.

_____, *The Coming Boom: Economic, Political, and Social,* New York: Simon & Schuster, 1983.

Klein, Gary, J. Orasanu, R. Calderwood, and C. E. Zsambok (eds.), *The Bottom Line: Naturalistic Decision Aiding,* Norwalk, NJ: Ablex Publishing, 1993.

Leedom, Dennis K., *Sensemaking Symposium,* Command and Control Research Program, Office of the Assistant Secretary of Defense for Command, Control, Communications and Intelligence, Washington, DC: Command and Control Research Program (CCRP), 2001. As of April 6, 2007:
http://www.dodccrp.org/events/2001_sensemaking_symposium/index.htm

Lempert, Robert J., "A New Decision Science for Complex Systems," *Proceedings of the National Academy of Sciences Colloquium*, Vol. 99, Suppl. 3, 2002.

Lempert, Robert J., David G. Groves, Steven W. Popper, and Steve C. Bankes, "A General Analytic Method for Generating Robust Strategies and Narrative Scenarios," *Management Science*, April 2006.

Lempert, Robert J., Steven W. Popper, and Steven C. Bankes, *Shaping the Next One Hundred Years: New Methods for Quantitative Long-Term Policy Analysis,* Santa Monica, CA: RAND Corporation, 2003. As of April 16, 2007:
http://www.rand.org/pubs/monograph_reports/MR1626/

MacKerrow, Ed, "Understanding Why: Dissecting Radical Islamist Terrorism with Agent-Based Simulation," *Los Alamos Science*, Vol. 28, 2003.

Millot, Dean, Roger C. Molander, Peter A. Wilson, and Marc Dean Millot, *The "Day After . . ." Study: Nuclear Proliferation in the Post-Cold War World—Volume II, Main Report,* Santa Monica, CA: RAND Corporation, MR-253-AF, 1993. As of April 16, 2007:
http://www.rand.org/pubs/monograph_reports/MR253/

Molander, Roger C., Peter A. Wilson, B. David Mussington, and Richard Mesic, *Strategic Information Warfare Rising,* Santa Monica, CA: RAND Corporation, MR-964-OSD, 1998. As of April 16, 2007:
http://www.rand.org/pubs/monograph_reports/MR964/

Morrison, Jeffrey G., *Multivariate Statistical Methods,* New York: McGraw-Hill, 1990.

Mussington, David, "The 'Day After' Methodology and National Security Analysis," in Stuart Johnson, Martin Libicki, and Gregory Treverton (eds.), *New Challenges, New Tools for Defense Decisionmaking*, Santa Monica, CA: RAND Corporation, 2003. As of April 16, 2007:
http://www.rand.org/pubs/monograph_reports/MR1576/

National Intelligence Council, *Global Trends 2015: A Dialog About the Future with Nongovernment Experts,* Washington, DC: Director of Central Intelligence, 2000.

_____, *Mapping the Global Future: Report on the National Intelligence Council's 2020 Project,* Washington, DC: Director of Central Intelligence, 2004.

Pate-Cornell, M. Elisabeth, and Seth D. Guikema, "Probabilistic Modeling of Terrorist Threats: A Systems Analysis Approach to Setting Priorities Among Countermeasures," *Military Operations Research*, Vol. 7, No. 4, December 2002, 5–20.

Prietula, M. J., K. M. Carley, and L. Glasser, *Simulating Organizations: Computational Models of Institutions and Groups*, Cambridge, MA: MIT Press, 1998.

Robalino, David A., and Robert J. Lempert, "Carrots and Sticks for New Technology: Crafting Greenhouse Gas Reduction Policies for a Heterogeneous and Uncertain World," *Integrated Assessment*, Vol. 1, 2000, 1–19.

Rosen, Julie A., and Wayne L. Smith, "Influencing Global Situations: A Collaborative Approach," *Chronicles Online Journal*, undated.

Rumsfeld, Donald, *Quadrennial Defense Review Report,* Washington, DC: Department of Defense, 2001.

_____, *Quadrennial Defense Review Report,* Washington, DC: Department of Defense, 2006.

Schoemaker, Paul, "Scenario Planning: A Tool for Strategic Thinking," *Sloan Management Review*, Winter 1995, 25–40.

Schwartz, Peter, *The Art of the Long View: Planning for the Future in an Uncertain World,* New York: Currency, 1995.

Simon, Herbert, *Sciences of the Artificial,* 2nd ed., Cambridge, MA: MIT Press, 1981.

_____, *Models of Bounded Rationality,* Vol. 1, Cambridge, MA: MIT Press, 1982.

Sisti, Alex F., and Steven D. Farr, "Model Abstraction Techniques: An Intuitive Overview." Last accessed in 2005:
http://www.rl.af.mil/tech/papers/ModSim/ModAb_Intuitive.html

Sterman, John D., *Business Dynamics: Systems Thinking and Modeling for a Complex World,* Boston, MA: McGraw-Hill/Irwin, 2000.

Szayna, Thomas S., *The Emergence of Peer Competitors: A Framework for Analysis,* Santa Monica, CA: RAND Corporation, 2001. As of April 16, 2007:
http://www.rand.org/pubs/monograph_reports/MR1346/

Tolk, Andreas, "Human Behaviour Representation: Recent Developments." Last accessed in 2001:
http://www.vmasc.odu.edu/pubs/tolk-human01.pdf

Tversky, Amos, and Daniel Kahneman, "Availability: A Heuristic for Judging Frequency and Probability," *Cognitive Psychology*, Vol. 5, 1973, 207–232.

Uhrmacher, Adelinde, Paul A. Fishwick, and Bernard Zeigler, "Special Issue—Agents in Modeling and Simulation: Exploiting the Metaphor," *Proceedings of the IEEE*, Vol. 89, No. 2, 2001.

Uhrmacher, Adelinde, and William Swartout, "Agent-Oriented Simulation," in Mohammad S. Obaidet and Georgios Papadimitrious (eds.), *Applied System Simulation*, Dordrecht, Netherlands: Kluwer Academic, 2003, 239–259.

van der Werff, Terry J., "Scneario-Based Decision Making—Technique." Last accessed in 2000:
http://www.globalfuture.com/scenario1.htm

Wack, Pierre, "Scenarios: Uncharted Waters Ahead," *Harvard Business Review*, 1985.

Wagenhals, Lee, Insub Shin, and Alexander E. Levis, "Executable Models of Influence Nets Using Design/CPN," *Proceedings of Workshop on Practical Uses of Colored Petri Nets and Design/CPN*, DAIMI PB-532, Aarhus University, Denmark, June 1998.

Wolf, Charles, Jr., K. C. Yeh, Benjamin Zycher, Nicholas Eberstadt, and Sungho Lee, *Fault Lines in China's Economic Terrain,* Santa Monica, CA: RAND Corporation, 2003. As of April 16, 2007:
http://www.rand.org/pubs/monograph_reports/MR1686/